• • SCOUTING AND • •
RECONNAISSANCE
IN SAVAGE COUNTRIES

• • SCOUTING AND • • RECONNAISSANCE IN SAVAGE COUNTRIES

BY

CAPTAIN C.H. STIGAND
F.R.G.S., F.Z.S.
"The Queen's Own" Royal West Kent Regiment
Author of "Central African Game and its Spoor"

Introduced by
C.A. BROWN

Original Edition
𝔏𝔬𝔫𝔡𝔬𝔫
HUGH REES, Ltd., 119, PALL MALL, S.W.
1907

This Edition
𝔊𝔬𝔲𝔩𝔟𝔲𝔯𝔫
ROPESEND CREEK PRESS
2019

This edition of

SCOUTING & RECONNAISSANCE IN SAVAGE COUNTRIES

© 2019 ROPESEND CREEK PRESS

All rights reserved. No part of this edition of Stigand's *Scouting & Reconnaissance in Savage Countries* may be reproduced in any form by any electronic or mechanical means including photographing, photocopying, recording, or information storage and retrieval without permission in writing from the publisher.

ISBN-13: 978-0-9944376-4-8
ISBN-10: 0-9944376-4-1

Since 2012, Ropesend Creek Press has specialised in bringing selected rare and classic works back into print for discerning outdoorsfolk.
Search "ROPESEND CREEK PRESS" for further great titles from our catalogue.

INTRODUCTION

TODAY, toward the end of the second decade of the 21st Century, one could be forgiven for not immediately recognising the name Chauncey Hugh Stigand.

C.H. Stigand

Unlike his fellow African explorer, hunter, soldier and naturalist, Frederick Courteney Selous, Stigand lived his remarkable life with a healthy disdain for self-aggrandisement.

His good friend and fellow adventurer

Theodore Roosevelt wrote of Stigand[1]:

> He tells too little of his own achievements. He has, as I can myself testify, the reputation among all firstclass African hunters of being himself one of the foremost. He is equally fond of venturing into unknown regions and of the chase of dangerous game, and is an adept in the especially difiicult art of wood and bush tracking and stalking. Three times he has been nearly killed by his quarry: once by a rhinoceros, once by a lion, and once by an elephant. It is unfortunate that he will not give us more minute and extended accounts of his own personal adventures — it is as difficult to get Captain Stigand to tell what he has himself done as it was to get General Grant to talk about his battles.
>
> After this manuscript was in my hands, Captain Stigand was nearly killed by an elephant. It was in the Lado, and he was taken down to Khartoum ; but his letters to his friends at home touched so lightly on the subject that they had to obtain all real information from outside sources.

A true quiet achiever, Stigand was one of Africa's foremost hunters and naturalists and a prolific author with no fewer than 13 books to his credit

---

[1] Foreword to *Hunting the Elephant*, C.H. Stigand, The Macmillan Company, New York, 1913

published between the years 1906 and 1923 (the latter posthumously).

One of the largest "tall tales" about Stigand was that he escaped an attacking lion by beating it with his fists to such an extent that the lion was found dead the next morning. As with the best tall tales, this one has a healthy grain of truth. The 1905 incident was so astounding to his comrades, that one wrote to the *Daily Graphic* newspaper[2] in Nairobi, East Africa describing a hunt of some problem lions in a small outpost called Simba on the Uganda railway. With the name of "Simba" which literally means lion in the Swahili tongue, it's not surprising the area had quite a large population of the creatures. Since lions were a problem in the area, causing casualties among the local African population and their livestock, Stigand set out to reduce the lion population. Stigand waited in ambush on a tank stand near a water hole. Early in the night he shot a lioness, and later two large lions came sniffing around her remains. Stigand shot both. The anonymous correspondent continues the story:

> Seeing the beasts apparently dead, Captain Stigand descended from the water tank and walked towards the huddled brown mass. He was only a few feet off when the worst happened. The beast rose from the grass and sprang.
> With a mighty roar he leapt into the

---

2   *Officer's Thrilling Experience*, Daily Graphic, via *The Clarence River Advocate*, Tuesday, 5th of December, 1905.

air and the whole surroundings seemed hidden by his massive frame. The sight was truly a terrific one. Every hair in the lion's body stood out, and every vein swelled with fierce anger. A shot was quickly fired, but this only increased the ferocity of the attack, and then came a scene, the likes of which have rarely been enacted.

The lion seized the left arm of the hunter, and the man and beast rolled over together. With his right arm free the gallant soldier caught his feline assailant by the throat and fighting for his life, he struck the brute several times in a fantastic display of desperate pugilism. Once again they rolled over, the lion for the moment on top, and then the man, and so the fierce fight went on.

At last there was an unexpected lull. The lion, sick and wounded unto death, savagely shook his victim, and then to Captain Stigand's amazement, slunk off.

At the moment of writing, Captain Stigand lies in the Nairobi hospital where he is receiving unremitting attention, and where it is sincerely hoped he will recover from his terrible experience. It is interesting to note that three dead lions were found at the break of dawn the following day, and that predations upon the local populace and their cattle have apparently ceased.

In October of 1919, in his capacity as Governor of the Sudanese province of Mongalla, Stigand embarked upon an armed expedition against troublesome Alaib Dinka tribesmen who had attacked a police post and killed eight policemen. On the 8th of Decenber 1919, Stigand's column was being harried by Aliab Dinka fighters and found itself ambushed in long grass by several hundred. Stigand, LtCol White and 25 soldiers were speared to death.

Of all of the titles authored by Chauncey Hugh Stigand, *Scouting & Reconnaissance in Savage Countries* is one of the rarest today. A check at time of writing this introduction in January 2019 reveals only 13 copies extant in libraries worldwide. Published in 1907, the book quickly became a best seller, but only that single edition was ever printed.

A review of the book appearing in *The Geographical Journal*[3] described it thusly:

A POCKET-BOOK FOR TRAVELLERS.

'*Scouting and Reconnaissance in Savage Countries*'. By Captain C. H. Stigand, F.R.G.S., F.Z.S. London: Hugh Rees, Ltd. 1907. *Price 5s. net.*

Under the above title, Captain C. H. Stigand has produced a little handbook, which gives a great deal of information of a very practical nature in connection with matters that cannot fail to be of importance, not only to the scout in a hostile country, but to all pioneers and explorers. To begin with, there are useful

---

[3] *The Geographical Journal*, Vol. 30, No. 6 (Dec., 1907), p. 655,

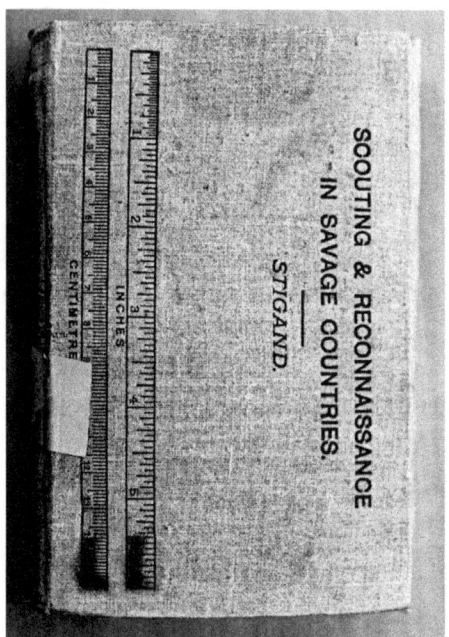

The cover of the original 1907 edition of *Scouting & Reconnaissance in Savage Countries* in the Editor's collection which was used to create this new paperback edition.

hints on finding the time and direction by means of rough bearings of the sun, stars, and moon, which, though not pretending to scientific accuracy, should be of assistance to any traveller unprovided with instruments who may be hurrying through a wild country. This is followed by a chapter on " Landmarks and General Information," another on " Tracking," which contains many good tips for following up and recognizing the spoor of various animals and men. After this follows a chapter of

" General Hints," one on " Tribal Customs and Differences," followed by one on " Reconnoitring Hostile Kraals or Villages."

There are also two appendices, the first of which gives instructions for using the star maps which are placed in pockets at the beginning and end of the book, while the second suggest various "'Exercises for Scouts," to assist in observing and noting events, and facts that may prove of importance later on.

Among the books consulted, Captain Stigand specially mentions this Society's '*Hints to Travellers*', and for more exact information on all the astronomical

This photograph of an original 1907 edition of Stigand's *Scouting & Reconnaissance in Savage Countries* from the Editor's collection illustrates nicely the wrap-around waxed linen cover.

> matters dealt with this should be referred to. The little work is strongly bound in pocket-book form, and at the end contains blank pages for notes and for keeping a rough field book.

In its original form, the book is pocket-sized and has a water-resistant waxed linen wrap-around cover which is closed with a press stud. inside the front cover, close to the spine is a pencil pocket, complete with a small eraser-tipped pencil. The inside of the rear cover has a pocket in which may be found three folded star charts, printed on linen.

The book's page count runs to around 140 pages of written content, and it is followed by 50 blank gridded pages and 50 blank field-book pages.

A completely practical publication devised by a completely practical man, the book was meant to travel in the pocket in the field and the waxed covers meant it was well protected from any sudden downpour. The blank pages gave the book immense utility as a field book for the explorer, hunter or naturalist.

The most astounding feature of the book was the three star charts which were found in a pocket inside the back cover. These charts were printed on linen, and like the book's cover, were waxed. The result was a tough, waterproof map which, when folded, took up very little space.

This orginal format, while immensely practical for the field, meant that the book is extremely difficult to reproduce in a normal book format. The length of the Equatorial Star chart alone is a couple of feet when the linen is unfolded, so some thought, and some compromises, have been necessary in order to publish this new edition.

Aside from the cloth wrap-around covers, the biggest difference between this new edition and the 1907 original is that the star charts have been incorporated into the body of the book as illustrations in a completely new Appendix (See *Appendix III.*). These charts may appear to be a little clunky to use, but they will work. The use of a small magnifying glass is recommended to read the data on the Northern Hemisphere and Southern Hemisphere charts in particular.

In the interests of brevity, we have not included the 50 blank gridded and 50 blank field-book pages in this edition.

Despite these differences, the book is extremely readable and will no doubt be a welcome addition to the research library of serious outdoorsfolk and particularly those with an interest in bushcraft, tracking and classic camping and expeditioning. The book is destined to become an instant classic - some 112 years after the its first publication!

At time of writing, plans are afoot to release a limited edition perfect replica of the original book, complete with waxed linen wrap-around cover, fold out illustrations, pencil and linen star charts. However, since this perfect replica edition requires a substantial investment in development, time and funds, for the time being, this new paperback edition is the only game in town.

2019 marks 100 years since the untimely death of Chauncey Hugh Stigand, and we at Ropesend Creek Press are proud and honoured at this time to be able to bring this rare and enigmatic little work out of obscurity and into the awareness of a much wider audience.

C. A. BROWN.

GOULBURN,
January 6th, 2019.

PREFACE

THE first time one accustomed only to civilised countries finds himself in the middle of a desert, he is apt to feel a sense of bewilderment and helplessness. He has a vague idea that he ought to steer by the sun and landmarks, but how to do this is quite a different matter. He has been brought up to hear that the sun rises in the east and sets in the west. This is something definite, but on looking up he sees the sun overhead —a great, hot, red staring ball in a zone of blue, but from which particular, part of the horizon it has sprung to reach this position, or to which part it intends to go down appears impossible to say.

As to landmarks, he remembers passing a little limestone hill, which he particularly noticed as he rode past, but is aghast to find that there are some twenty or thirty of these, all alike, but none just like the one passed, as he did not turn to look at its other side after passing.

Next, as to finding the way back to a camp or water. In a civilised country there is always a long, straight bit of rail, or road, or telegraph wire which you must strike, even if you go many degrees out of your course. A spot is almost unconsciously marked down by striking for it, leaving a road or hedge at a particular place at a particular angle; but both hedge and road are artificial, and in a country where there is no such thing as a straight line an entirely new and more complicated system of marking spots must be resorted to.

If one had an instructor to take one in hand, and by practical experience show one exactly how to set about learning "knowledge of country," one would soon gain the confidence of the old hand. We had no such instructor, nor did we know of any book that could help us with rough and elementary rules, so we had to pick up these rules by hard and tedious experience.

It is now hoped that these few hints, gained by much losing of camps, missing of water, retracing of steps, and wanderings about, may make the path of other beginners like ourselves a little softer. If our little book does this, all that we have hoped to do will have been accomplished.

It might be mentioned that calculations of bearings could be worked out with far greater accuracy by applying to The Nautical Almanac and Hints to Travellers. However, the ordinary man, and especially the soldier scout, does not wish to burden himself with scientific books and instruments, and a knowledge of logarithms and mathematics, so as to be able to work out to the tenth part of a second a bearing which he has no intention or ability to march on with less than two or three degrees minimum of error. We have done our best to work out our tables so that they should seldom give an error of more than two degrees, and in nine cases out of ten they will be within one degree.

We should like here to record our great indebtedness to Colonel Cholmeley Harrison, Major C. F. Close, C.M.G., Royal Engineers, and Captain H. W. B. Thorp, King's Own Yorkshire Light Infantry, for the exceedingly kind help they have afforded us, and for which we offer them our best thanks.. The chief books consulted were *The Nautical Almanac*, *Hints to Travellers*, and Burdwoods' *Tables*. We have assumed that our reader is acquainted with the *Manual of Map-Reading and Field-Sketching*.

C. H. STIGAND.

Zanzibar,
November 28th, 1906.

CONTENTS

INTRODUCTION .... *V*
PREFACE .... *XV*

I. KNOWLEDGE OF COUNTRY 1
(BEARINGS AND TIME.)

II. KNOWLEDGE OF COUNTRY 35
(LANDMARKS AND GENERAL OBSERVATIONS.)

III. TRACKING 55

IV. GENERAL HINTS 71

V. TRIBAL CUSTOMS AND DIFFERENCES 87

VI. RECONNOITRING HOSTILE KRAALS
    OR VILLAGES 95

APPENDIX I. HOW TO USE THE STAR CHARTS 105
APPENDIX II. EXERCISES FOR SCOUTS 109
APPENDIX III. STAR CHARTS 115
    (1) NORTHERN CONSTELLATIONS. 117
    (2) SOUTHERN CONSTELLATIONS. 121
    (3) EQUATORIAL STARS. 125

Scouting and Reconnaissance
in Savage Countries.

CHAPTER I.

KNOWLEDGE OF COUNTRY.
(Bearings and Time.)

Knowledge of country is the appreciation and committing to memory of natural features, and from this the ability to find one's way about. The man who has been brought up in a civilised country with towns and villages closely packed together, roads and rails, sign-posts and mile-stones, fields and hedgerows, cannot have the faintest conception of the difficulties of finding his way about a wild country.

At his first attempt they may appear so insurmountable that he is apt to give up trying and rely wholly on the native inhabitant to guide him from place to place. If, however, he will study a few elementary rules he will find the difficulties by no means so hopelessly perplexing as they may at first appear.

Finding one's way about a wild country consists of a knowledge of which direction to move in and for what distance to move in that direction.

This direction is ascertained either by means of bearings or by watching certain objects on the earth's surface (i.e. landmarks).

The distance may be judged either by eye or by landmarks or by time.

Bearings are obtained either from a compass, the sun, moon, stars, or wind.

The landmarks may be prominent objects such as hills, rocks, tall trees, or they may be less obvious objects such as river beds, patches or belts of certain kinds of trees, outcrops of certain kinds of rock, certain kinds of soil or grasses, rises and falls in the country, etc.

Time is judged by watch, sun, moon, or stars. In flat close country and in open deserts one is guided almost entirely by bearings and by knowledge of trees. In hilly country to carefully watch landmarks is nearly all that is required.

The method of marching on compass bearings is too well known to merit description, so we will here only deal with—

1. Bearings from sun.
2. Bearings from moon.
3. Bearings from stars.
4. Bearings from wind.

5. Time by sun.
6. Time by moon.
7. Time by stars.
8. Landmarks.

1. Bearings from Sun.

It is quite possible that the scout's compass may be lost or broken, or in many cases it may be inconvenient or impossible to use it, as when riding, when time is pressing, where there is iron in the soil, at night, etc. It is then necessary to be able to get bearings from the sun or stars.

When the principles guiding the movements of these bodies have been mastered, it will be found that the compass will seldom be looked at for rough bearings, it being so much quicker to look at the sun or stars, or one's own shadow, all of which can be done without having to stop.

We will first try and describe the movements of the sun and then endeavour to show how these may be turned to account.

Before proceeding, however, lest the reader should grow discouraged with the seeming intricacies of the different bearings at the different times of day, we would impress on him that the more practical experience he gains in the field the less important will the theoretical part appear to him. He will learn to always appreciate quickly and readily which direction is the north and in which direction he is to steer with regard to it.

Owing to the angle at which the earth's axis is turned towards the sun, the sun at different times of the year appears at noon to pass vertically above different places on the earth's surface. On June 22nd the sun passes vertically over lat. 23 ½° N. It then moves slowly southwards passing over the equator, till on December 22nd it is vertically above lat. 23 ½° S. Here it turns and slowly passes northwards again. In no latitude farther north than 23 ½° N., or farther south than 23 ½° S. is the sun ever vertically overhead. Between these two parallels are the tropics.

The "declination" of the sun is the amount of degrees that the sun is from the equator at noon, or, in other words, that latitude in which it is vertically overhead at noon. When the sun is vertically above the equator, the declination is 0°. When the sun is vertically above lat. 23 ½° N., the declination is 23 ½° N.

The accompanying table shows roughly the declination of the sun on the 1st of every month throughout the year. From these any day in the year may be approximately ascertained with sufficient accuracy for our purpose. Should more accurate information be required *The Nautical Almanac* for the current year may be referred to.

Approximate Table of the Declination of the Sun for the 1st of every Month.

1st January,	S. 23°.
1st February,	S. 17°.
1st March,	S. 8°.
1st April,	N. 4°.
1st May,	N. 15°.
1st June,	N. 22°.
1st July,	N. 23°.
1st August,	N. 18°.
1st September,	N. 8°.
1st October,	S. 3°.
1st November,	S. 14°.
1st December,	S. 22°.

N.B. - It will be noticed that the second six months are practically a repetition of the first with opposite signs of north and south. To know how far north or south of you the sun is, it is necessary to know two things ; the first is your latitude, and the second is the declination of the sun for the given time of year.

To obtain the number of degrees the sun is from you : If the declination and your latitude are both north or both south, you subtract one from the other ; but if one is north and the other south, you add the two together.

Example.—On January 1st, in lat. 70° N., you wish to know where the sun will be at noon. Looking at the table, you see that the declination is

23° S. on January 1st. You are 7° N. Therefore you add the two, which makes the sun 30° S. of you. This means that at noon a line drawn from you to the sun is pointing south and makes an angle of 30° with the vertical, and hence 60° with the horizontal. If, on the other hand, your latitude had been 7° S., this angle would only have been 16°; but it would still be to the south, as the sun is at 23° S., which is farther south than your latitude.

Conversely it is by the exact measurement of this angle of the height of the sun at noon that geographers are able to fix the latitude of a place.

For the purpose of the scout it is only necessary to know the approximate latitude, which he can obtain from a map, and from this the approximate angle of the sun. It is hardly likely that the scout would ever find himself in a place the approximate latitude of which he did not know within a degree or so; but it might be useful to remember that in the northern hemisphere the angle of the height of the Pole star above the horizontal, measured with sextant, clinometer, or Abney's level, is approximately the latitude of the place from which such observation is made.

For bearings it is really only necessary to know whether the sun is north or south of you at noon, but knowing the exact amount it is to the north or south enables you to determine when it is noon. (See Time by Sun.)

In addition to this yearly movement of the sun north and south, there is a diurnal movement.

The sun rises due east at the equinoxes, and at this season at the equator moves straight up the heavens, till it is vertically overhead, and straight down the other side, setting due west. From any other part of the earth's surface north or south of the equator, at the equinox, the sun would appear to rise due east, and move across the south or north side of the heavens, as the case may be, setting due west.

The equinoxes are when the declination of the sun is 0°, and occur about March 21st and September 23rd. On these dates the sun, as we have said, rises due east and sets due west for all latitudes. At other times the bearing of the sun at rising and setting varies for different latitudes.

The greatest departure from true east and west is during the solstices—that is, when the sun is at its greatest northern or southern declination, about June 22nd and December 22nd. On June 22nd the sun has its greatest departure to the north at rising and setting, and on December 22nd its greatest departure to the south.

The accompanying table gives these departures for the different latitudes at the solstices, when they are greatest. The angle given is the amount the sun rises and sets north of true east and west on June 22nd and south of true east and west on December 22nd.

From this table the departures of the sun for other dates can be roughly worked out, as they vary roughly in proportion to the declination of the sun.

N.B.—These departures vary approximately

in proportion to the declination, not in proportion to the number of days from the solstice or equinox.

Bearings of Sun on Rising and Setting during the Solstices for Different Latitudes.

Equator.	Lat. 0°	23°.
	Lat. 10°	24°.
	Lat. 20°	25°
	Lat. 30°	27°
	Lat. 40°	31°
	Lat. 50°	38°
	Lat. 60°	53°

Above bearings to be read north of east and west at northern solstice (June 22nd) and south of east and west at southern solstice (December 22nd).

N.B — in the tropics the departure from east and west of the sun at rising and setting can be taken as the same as the declination, which will be near enough for practical purposes.

In getting one's bearings from the sun it will be found more convenient to watch shadows instead of looking at the sun itself, which is tiring to the eyes. It is generally one's own shadow which is observed, but that of any perpendicular object, such as a tree or stick, may be used.

From the tables given above we can now make the following rules :

If the sun is north of your latitude, your shadow will point approximately due south at mean noon ; and if the sun is south of your latitude, your shadow will point due north at noon.

In Great Britain, which is between lats. 50°N. and 58° N., the sun is always south because it never comes farther than 23 ½° N., and so your shadow at noon always points north.

Outside the tropics the sun at noon is always south of you in the northern hemisphere and north of you in the southern, the whole year round.

In the tropics the sun is sometimes north, and sometimes south, and twice a year overhead.

At the equinox your shadow always points west at sunrise and east at sunset.

At other times the position of the sun at rising and setting must be worked out from the declination and the table of departures given above and the bearings judged accordingly.

Let us take a simple example: You are in lat. 40°S. on May 12th and wish to know the bearings of the sun at different times of day. You must first find the declination for May 12th, so referring to the table of declinations you find that for May 1st it is 15° N. and for June 1st it is 22° N. Therefore on May 12th it will be about 18°N. Next you refer to the table of departures, and find that the greatest departure for lat. 40° is 31°. It departs 31° for a declination of 23 ½°, so for a declination of 18° it will have a departure of about 24°. As the declination is north this departure will be to the north of east at rising and to the north of west at setting, or, roughly, the sun will rise ENE. and set WNW. As we are in the southern hemisphere, the sun is due north at noon.

If sunrise is at 7 o'clock we have :

Sun rising	ENE.
Sun at 8.40 p.m.	NE.
Sun at 10.20 a.m.	NNE.
Sun at Noon	N.
Sun at 1.40 p.m.	NNW.
Sun at 3.20 p.m.	NW.
Sun at 5 p.m. (sunset)	WNW.

Having obtained the bearings of the sun at different hours, the hours themselves can be told by a watch, or, failing this, by the length of your shadow. (See "Time by Sun.")

The important bearings to know are :

(1) Whether the sun is north or south at noon.

(2) Bearing of the sun when rising and setting.

Once these are known, experience will teach the scout how much to swing round with regard to his shadow as the day advances without worrying about the other bearings.

He will also learn to judge roughly what length of shadow denotes a given bearing and given time of day. The great point is to always have in mind the direction of north, and the direction you wish to go with regard to it.

Example.—In lat. 10° N. you wish to march due west on November 7th.

Declination November 1st—14° S., so on November 7th will be about 16° S. Your latitude is 10° N., so the sun is 26° S. of you. Your shadow at noon will point north.

You are close to the equator, so the departure can be taken as the same as the declination. So sun will rise 16° S. of east, and set 16° S. of west.

At sunrise your shadow will point 16° N. of west, so you commence marching at an angle of about 16° to left of your shadow. As the day wears on you leave your shadow more and more to your right, till at noon you are marching at right angles to it.

If the march is continued in the afternoon you let your shadow gradually drop over the right shoulder, and as it will then be difficult to see, you will probably observe the shadows of trees as you pass them till near sunset, when you will be marching a little to the right of the sun.

The most difficult time of day is at and near noon, when shadows are shortest, and when an hour in time does not make much corresponding increase or decrease in the length of shadow.

It must be remembered, however, that the sun is not the only guide, for there is the wind, as also the signs of prevailing winds, landmarks, etc., to go by, and from which the observations of the sun may be checked. Whenever possible, points must be marked down ahead in the right direction, to use when they are visible, both in checking your direction with the sun or to use alone when the sun goes in.

2. Bearings from Moon.

The movements of the moon are so intricate that one cannot lay down any definite rules for guidance. However, the moon is of the greatest service at night, and although one may be unaware of its exact bearing, watching its position may be of the greatest help to one in finding the way back when out at night.

The moon goes through in one month much the same process as the sun does in a year, so its place of rising and setting varies considerably from day to, day. However, one should generally, and if outside the tropics always, know if it is north or south of you when at its zenith (highest point).

Its bearing at rising might be taken with a compass, or noted by some star it rises near, and from this its declination can be found out by reversing the tables. Once its declination is known it can be treated (for that night at least) just as if it were the sun with a similar declination. As it gives an appreciable shadow, one has also this to guide one as to its height. The declination can always be found out by reference to *The Nautical Almanac*.

Twice every month the moon must rise due east and set due west, which is useful to remember.

3. Bearings from Stars

The heavens revolve about two points, called the north and south poles of the heavens.

North can be obtained from the Pole star, which is near enough to the north pole to be sufficiently accurate for our purpose.

This star can be located by means of the two pointers of the Great Bear or " Charles's Wain," for a prolongation of the line joining them almost touches the Pole star.

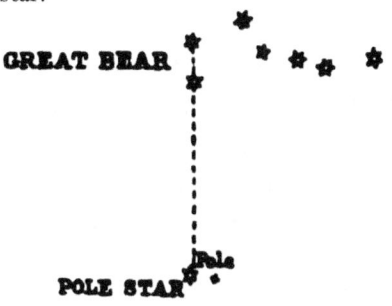

When the Pole star is below the horizon, and the pointers are visible, if the latter are vertically above one another as in diagram they can be taken as north, while at other times the direction of the north point can be judged from the direction in which they are pointing. The distance between the Pole star and the pointers should be often observed, so as to accustom the eye to estimate this distance.

The height of the Pole star above the horizon corresponds to your latitude N., so if the latitude is unknown it may be roughly estimated by the altitude of this star (sec page 7), whereas if you know your latitude it may help you to find this star on a cloudy night, when few stars are visible, as you will then know at about what height above the horizon to look for it. The Pole star can only be seen in the northern hemisphere ; and on passing southwards of lat. 10° N., owing to its proximity to the horizon it will become invisible sooner or later, according to the state of the atmosphere. South of the equator it will be below the horizon. The pointers, however, may be used some 20° or 30° farther south for obtaining the direction of north.

At about lat. 20° N. the Southern Cross comes into view, and anywhere south of this latitude may be used for obtaining south. When the Cross is vertical, its longer diagonal can be taken as south. At all other times if this longer diagonal is produced three and a half times its own length, a point near the south pole is reached.

The Cross must not be confused with the false cross, which is in Argo, and is larger in size.

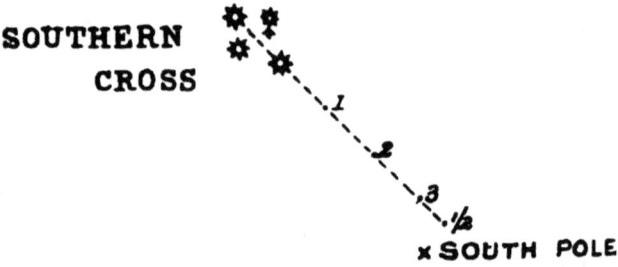

South may also be obtained approximately from Alpha and Beta Centauri, the pointers to the Southern Cross. When these two stars are horizontal, south falls nearly midway between them.

Other useful stars are those of Orion, on the celestial equator, which rise and set due east and west for all latitudes. Other stars vary in their places of rising and setting for different latitudes.

The accompanying table gives the most important stars and their declinations ; these declinations correspond to the latitudes in which they pass vertically overhead. As such they will be found useful later on in computing time at night.

The places of rising and setting can be ascertained from the declinations.

Table of the Declinations of the Principal Stars.

NORTHERN HEMISPHERE.

α Ursae Majoris (Dubhe)	62° N.
γ Ursae Majoris	54° N.
η Ursae Majoris	49° N.
α Aurigae (Capella)	45° N.
β Perseus (Algol)	45° N.
α Lyrae (Vega)	38° N.
α Geminorum (Castor)	32° N.
β Geminorum (Pollux)	28° N.
The Pleiades	24° N.
α Arietis	23° N.
α Bootis (Arcturus)	20° N.
α Tauri (Aldebaran)	16° N.
β Leonis (Denebola)	15° N.
α Leonis (Regulus)	12° N.

α Aquilæ (Altair)	9° N.
α Orionis (Betelgeux)	7° N.
γ Orionis (Bellatrix)	7° N.
α Canis Minoris (Procyon)	5° N.
δ Orionis	Equator.

SOUTHERN HEMISPHERE.

β Orionis (Rigel)	8° S.
α Virginis (Spica)	11° S.
α Canis Majoris (Sirius)	17° S.
β Scorpii	20° S.
δ Canis Majoris	26° S.
α Scorpii (Antares)	26° S.
ε Canis Majoris	29° S.
α Piscis Australis (Femalhaut)	30° S.
α Columbae	34° S.
ζ Argus	34° S.
α Phoenicis	43° S.
λ Argus	43° S.
α Gruis	47° S.
α Argus (Canopus)	53° S.
β Aræ	55° S.
γ Crucis	55° S.
α Pavonis	57° S.
δ Crucis	57° S.
β Crucis	57° S.
ι Argus	59° S.
η Argus	59° S.
α Centaurus	60° S.

β Centaurus	60° S.
α Crucis	62° S.
α Trianguli Australis	69° S.

The places at which these stars rise and set for various latitudes can be found out from the following data, but it must be remembered that no star is to be used whose declination added to your latitude makes more than 80°, as the following data will not apply to them ; e.g. in lat. 35° N. or S. the farthest star which can be used to the north is Capella, as its declination 450 added to the latitude 35° makes just 80°.

Rules for finding the Places of Rising and Setting of Stars from their Declination.
At the Equator: All stars rise and set north and south of true east and west the amount of their declination. Thus Capella observed at the equator rises NE. and sets NW., and Trianguli rises SSE. and sets SSW.

Roughly the same holds good for anywhere in the tropics, but for greater accuracy—

In Lat. 10°: Add 1° for every 20° of star's declination.

In Lat. 20°: Add 1° for every 10° of star's declination.

In Lat. 30° : Add 2° for every 10° of star's declination.

In Lat. 40°: Add 40 for every 10° of star's declination.

In Lat. 50° : Add 7° for every 10° of star's

declination.

In Lat. 60°: For stars close to celestial equator double the declination.

Examples—

(1) Bearings of Vega required in lat. 40°.

Declination of Vega is 38° N.

Add 4° for every 10°, which makes nearly 16°, therefore Vega rises 54° to the N. of east.

(2) You are in lat. 30°, and wish to know which stars may be useful to you, besides the Pole star or the Southern Cross and equatorial stars. For lat. 30° you have to add 2° for every 10° of declination.

Looking at the tables Vega is 38° N., add 8°, which makes 46° N., which is practically NE. rising and NW. setting.

Arcturus is 20° N., add 4°, which brings it to 24° N., or nearly ENE. at rising and WNW. at setting.

β Scorpii is likewise ESE. and WSW.

ζ Argus will be 41° and λ Argus will be 51°. As they will rise near each other, to get SE. and SW. you can steer between the two.

If none of these stars are on the horizon, you know from their declination where they will be when highest, so you can estimate their intermediate positions in the same way as for the sun.

(3) In lat 60° it is required to march WSW. at night.

It is evident that a star setting 22 ½° S. of west

would be the best guide obtainable.

As declinations are doubled for lat. 60° a star whose declination was half this amount, *viz.* 11°, would be such a one. Looking at the tables we see that Spica has this declination. If Spica is nearly setting at the time we wish to march, we would use it, otherwise we must choose another star and march at an angle to it.

Learn these stars, which will be useful to you in the latitude which you are in. When a star rises too high to be of use or is about to set, any small star immediately below or above can be chosen by eye till another point is recognised. However, a star chosen in this way should not be used for long unless its bearing at rising and setting and declination is known, or its lateral movement can be judged. Now as to the estimating of this lateral movement : A star is always moving from east to west, but the rate at which it moves varies for different declinations and different latitudes. You can find by the rules above its place of rising and setting. Therefore, unless it passes vertically overhead, it moves through the angle between those points in the time between its rising and setting.

For instance, if a star rises and sets northeast and north-west, it must pass through an angle of 90° during the time that it is above the horizon. It is then necessary to roughly estimate the length of time that it is above the horizon, so as to obtain the number of degrees it moves from east to west per hour. It would be difficult to lay down any simple rules for estimating this length of time, but it may be roughly judged from the following data:

Stars vary as to the length of time that they are above the horizon in exactly the same way as does the sun for a corresponding declination.

If you study "Time by Sun," and the table, it will give you a rough idea. That is to say, equatorial stars correspond to the sun on the equator, and are twelve hours from rising to setting all the world over. When you are at or near the equator, nearly all stars are about twelve hours from rising to setting. Stars north and south of the equator are longer above the horizon the nearer you are to the pole, and the nearer they are to the pole in your hemisphere and shorter, the nearer they are to the opposite pole. Thus a star with declination 23 ½° will be the same length of time above the horizon as the sun at the solstices—that is, if the star and the observer are both north or both south, take twice the number of hours in the table, which will give the number of hours from rising to setting ; but if one is north and one south, subtract this result from 24, and you get the number of hours of a star in the opposite hemisphere. Do not choose a star which passes overhead, or nearly overhead, for this. We will touch on those presently.

Example.—In lat. 30° S. you wish to know the lateral movement of Arcturus. Its declination is 20° N. So for lat. 30° it rises 24° to N. of east.

It is north of you at its highest point. It moves from 24° N. of east to 24° N. of west during the time that it is above the horizon—that is, it goes through an angle of 132°.

Sun for lat. 30° and declination 23 ½° shows 6.58. Therefore Arcturus will be a little less, say 6.45. Double this is 13½ hrs. As Arcturus is in the opposite hemisphere, you subtract from 24 : result, 10½ hrs. Divide 132° by 10½, and you get 12°, which will be the lateral movement of the star per hour approximately. You should not use very northern or southern stars like this, but if you want to march northwards or southwards try to use the Pole star or the Southern Cross.

As a rough rule it can be taken that when selecting a star by compass bearing or from a star chart, between the polar and equatorial stars, for a night march, you should choose a star about 5° to east of required bearing and march on it for about one hour, when it should have moved about 10° to west, and so you will have kept an average bearing of that required.

For stars that pass overhead, find out where they rise and where they are at their highest point; take the difference between this and divide it by half the number of hours it is above the horizon.

Example.—In lat. 30° N. you want lateral movement of ε Orionis. It is on the equator, therefore it rises due east, and at the highest point is 30° south of you. Being on the equator, it takes twelve hours from rising to setting, and thus six hours to reach its highest point. Therefore, it takes six hours to travel 30° to south, and another six hours to come back again to west So ascending it is moving south at the rate of 5° per hour, and descending it is moving back to north at the same rate.

4. Bearings from Wind.

Marching across flat, closed country, without footpaths, direction has to be maintained on a cloudy day in a great measure by the wind and by the signs of prevailing winds. On dark cloudy nights wind is often the only indication obtainable of the direction in which to travel.

Where there are hills the wind is liable to local variations, but in flat country, whether wooded or desert, the wind is generally fairly constant.

As it will only be in flat and featureless countries that one will ever have to depend on wind for bearings for any length of time, the local variations in wind common to hills and valleys are the less important.

In most countries, and especially in countries where there are large flat expanses, the wind blows very steadily in one direction through a great part of the year, and during the remaining part it may blow from another quarter, with seasons of variable winds between the changes. Thus over the greater part of South and East Africa there is a steady wind from the south-east blowing all the dry weather.

In most parts of the world these winds, which return about the same time yearly, are well known and have names, eg. trade winds, monsoons, kharif, sirocco, kusi, etc. If their direction is known it is easy enough to steer an approximate bearing by their help, and during the seasons when they are not blowing by the traces they have left behind them, or by other winds blowing steadily.

The direction of the wind may be ascertained by watching the bending and waving of grass and boughs, by filtering sand or dust through the fingers to observe in which way it blows, or by moistening a finger and holding it above the head, when the side which feels cold will be to windward, or by smoking.

It must be remembered for any of the latter that one should be standing still, or the true direction of the wind will not be obtained. Of course only a rough approximation can be made by any of the above methods, but good enough for maintaining a rough bearing. One of the best ways of noting the direction of the wind is perhaps by watching the drift of the smoke from the camp fire.

Hunters after elephant or buffalo, when the direction of the wind has to be known to a nicety, often take a small bag of flour with them, which they shake at intervals, observing the direction taken by the powdered flour. Where the ground is dry, it is only necessary to kick up a little dust or kick an ant-hill as one passes it, to obtain the same result.

A strong wind can be felt either in front, blowing on one or the other cheek or ear, or at the back of the head, as the direction requires.

On the sea coast there are also sea and land breezes, the wind blowing landwards during the nights and seawards during the day.

It should be noticed whether the prevailing winds are characteristic of a wet or a dry season, as the signs they will leave will differ accordingly. Such signs are of the utmost importance and include :

In wooded country, which side of trees are weather-beaten.

Rocky country, which side of rocks and stones are most covered with moss' and lichen (the bare side is that exposed to wind).

Sandy country, in which direction the drifts are piled. Sand drifts are steep on the lee side and rounded on the weather side. In the Great Nefud of Central Arabia these drifts are said to run very regularly nearly east and west.

Sand piled to leeward of rocks, stones, and trees.

Which side of ponds and lakes are most washed by waves.

In which direction grass is bent.

During the south-east wind in Central Africa, the grass dries and is burnt, and the charred stalks can be seen broken down and pointing uniformly north-west.

5. Time by the Sun.

Time is told from the sun chiefly by the length of ones shadow.

In damp, hot climates watches are constantly going wrong, and in wild countries there is no means of getting them repaired. Hence the value of being independent of a watch.

It is also necessary to be able to tell the time by the sun when marching, as you must steer at different angles to your shadow according to the time of day—or, in other words, you must make different angles with your shadow according to the length of that shadow.

There is a slight difference between apparent time (time as told by the sun) and mean time (time as told by watches), and twice a year they differ as much as sixteen minutes. The reason for this is that the time between the sun reaching its highest point on two succeeding days is not always exactly twenty-four hours, but is sometimes a little more and sometimes a little less. However, if the scout has a watch, he will generally set it by sunrise in an uncivilised country, and so in either case he will be keeping to apparent time, and need not worry himself about the difference between that and mean time.

It might be useful to know that apparent and mean time coincide four times a year— *viz.* about April 16th, June 14th, September 1st, and December 25th—so a watch set by the sun about these dates would be also set to mean time.

At the equinox the sun rises and sets at six o'clock in all latitudes, and in the tropics this time can be taken as a rough rule for the whole year. The equinoxes are, as we have said, about March 21st and September 23rd.

North of the equator the longest day is June 22nd, and the shortest December 22nd.

South of the equator the exact reverse holds.

The time of the rising of the sun on the longest day is the time of its setting on the shortest. Twice the hour of rising is the length in hours of the night. Twice the hour of setting is the length in hours of the day.

The following table gives the time of the sun's rising on the shortest day and setting on the longest for various latitudes. From these the times of sunrise and sunset can be worked out for any day in the year or for any latitude, the times varying in the same proportion as does the declination of the sun.

Table giving Times of Sun's Rising on the Shortest Day and Times of Setting on the Longest Day for different Latitudes.

Equator	6.2	Lat. 40°	7.25
Lat. 10°	6.18	Lat. 50°	8.5
Lat. 20°	6.36	Lat. 60°	9.15
Lat. 30°	6.58		

Example.—Wanted, time of sunrise and sunset for March 1st in lat. 35° N. We are in northern hemisphere, so the longest day is June 22nd and shortest December 22nd.

Sun rises 6.58 for shortest day in lat. 30° and 7.25 in lat. 40°, so in lat. 35° N. it will rise 7.11 on December 22nd. It rises at 6 o'clock at the equinox when declination is 0° and 7.11 when declination is 23½° S. (*viz.* December 22nd). Therefore sun rises 1 hr. 11 mins. after 6 o'clock for declination of 23½° S.

Referring to table of declinations, we find that on March 1st it is 8°. If it is 71 mins, for 23½°, it will be 24 mins. for 8°, so the sun rises 24 mins. after six— *viz.* 6.24 a.m. —and therefore sets 5.36 p.m. This is, of course, apparent time, not mean time.

To set your watch by sunrise or sunset you must make allowance for anything, such as hills, trees, etc., blocking the view. It is only at sea and on absolutely flat open desert that the sun will be seen at the moment of rising.

The example above shows us how to get the times of sunrise and sunset. Noon is when the observer's shadow is shortest or non-existent; the other times of day may be estimated roughly as follows :

When the declination corresponds to one's latitude, if sunrise is at 6 a.m., then :

7 a.m. and 5 p.m. will be when the sun is about the length of the palm above the horizon— the arm held out at full length. A little practice will soon enable you to do this.

8 a.m. and 4 p.m. will be when the sun is

twice this distance above the horizon.

9 a.m. and 3 p.m. will be when one's shadow is equal to one's height.

10 a.m. and 2 p.m. will be when one's shadow is a little more than half one's height.

11 a.m. and 1 p.m. will be when one's shadow is a little less than half one's height.

Noon will be when there is no shadow.

6 p.m. will be sunset.

If sunrise is earlier or later than 6 a.m., these times will be correspondingly greater or less—
e.g.. if sunrise is at 5.30 a.m. sunset will be at 6.30 p.m., and a shadow equal to your height will
denote 8.45 a.m. and 3.15 p.m., and so on. Each division of time given above is a twelfth of the period between sunrise and sunset, or a sixth of that between sunrise and noon.

When the sun is not vertically above your latitude, your shadow will be correspondingly longer the farther the sun is from you.

By noting the difference day by day it will be found in time easy to judge.

The length of your noonday shadow can always be found by setting up a right-angled triangle as follows :

Example.—Wanted, length of noonday shadow on November 1st in lat. 16° N.

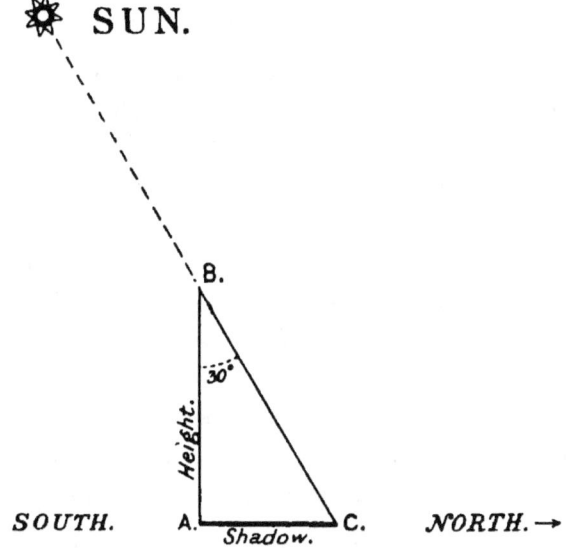

Declination for November 1st is 14° S., so the sun is 30° to south of you.

Set up right-angled triangle ABC. Where A B is your height and angle A B C is 30°, then A C will be the length of your noonday shadow, and, because the sun is south of you, will point north. For other times of day take the complement of A B C, which in this case will be 60°, and this will be the angle of the sun's height at noon. Then divide it up into portions to suit different hours of the day, and make right-angled triangles to ascertain the length of shadow for those hours. Thus if sunrise is at 6 a.m. we know, in above instance, height of sun (technically referred to as sun's altitude) at 7 a.m. and 5 p.m. will be 10°; at 8 a.m. and 4 p.m. will be 20°; at 9 a.m. and 3 p.m. will be 30°;

and so on.

Example.—To find length of shadow at 8 am. in above case set up A B equal to your height.

You then want angle A C B 20°, so you make the angle ABC the complement of 20°—*viz.* 70°—and then A C will be the length of your shadow.

N.B.—The complement of an angle for this purpose is obtained by subtracting it from 90°.

6. Time by the Moon.

As a rough rule it can be taken that the new moon rises at sunrise and sets at sunset.

The first quarter rises at noon and sets at sunrise.

Full moon rises at sunset and sets at sunrise.

The last quarter rises at midnight and sets at noon.

The moon rises roughly fifty minutes later every day.

The above will give some indication of how to gauge time at night, but it would be difficult to lay down any hard-and-fast rules on anything so complicated as the movements of the moon.

7. Time by the Stars.

About the end of March, the Southern Cross is standing upright at midnight, and on its side 6 hrs. before and 6 hrs. later. Therefore it will be slanting at an angle of 45°, to left and right respectively, at 9 p.m.

and 3 a.m.

At other times of year its position at sunset should be noted, and when it has revolved till it is at right angles to its first position you will know that 6 hrs. have elapsed. This means that it revolves like the hands of a watch, with the south pole as centre, and passes through an angle of $15°$ per hour. After a little practice it is not very hard to tell whether it has, say, revolved through an angle of $15°$ or $30°$—that is, whether a period of one or two hours has elapsed since you made your first observation.

The pointers a and β Ursae Majoris are upright at midnight at the end of February, and will similarly revolve through $90°$ and be lying on their side 6 hrs. later. If their position is carefully noted at sunset, the hours during the night can be judged by the amount they have revolved, in the same way as with the Cross.

Time can also be judged by the stars—to east at rising and west at setting; also by those which pass vertically overhead—i.e. those whose declinations correspond to the latitude.

When the places of the rising and setting and the declinations of stars are known, time at night may be estimated from the fact that those near the equator take 6 hrs. from rising to reach their highest point.

Example.—You are bivouacking in lat. $8°$ S., and at 9 p.m. you lie down to sleep and wish to start on again 3 hrs. before sunrise—that is, about 3 a.m.

Looking at the east side of the heavens you see Rigel about half-way up the heavens. You see by the table that it passes vertically overhead. At 3 a.m. it will be half-way down the west side of the heavens, so you lie down to sleep with your face in that direction, and, if possible, arrange to have a branch of a tree covering that point from where you lie. When Rigel reaches level with this branch, or you see that it is half-way down the west side of the sky, you know that it is time to start.

If you are judging time by a northern or a southern star, you will know that it has reached its highest point when it is vertically above north or south pole, or when it cuts an imaginary line drawn from the pole to a point vertically overhead ; this point is called the zenith.

Venus is a useful planet to observe, as never being far distant from the sun ; when seen in the night, it proclaims the near approach of dawn.

N.B.—Stars rise four minutes earlier every day.

8. Landmarks.

Before leaving the subject of bearings it should be stated that every effort should be made to supplement and check them by landmarks, as errors are bound to be made in working on rough bearings alone.

Bearings worked out from the tables should have no very serious error, but it will be in the endeavour to march on these bearings that error will creep in.

This can be minimised by practice and exercising care, and in a long march minor errors tend to counteract each other to a wonderful degree.

An error of $2°$ in marching twenty-five miles will give a lateral deviation of about one mile. It will seldom be necessary, even in thick country, to march as far as this without being able to observe a single landmark of any kind. A march of several days without appreciable landmarks might occur in some deserts, but in such country landmarks can be picked up at tremendous distances, and so an error of several miles can be corrected.

CHAPTER II.

KNOWLEDGE OF COUNTRY.
(LANDMARKS AND GENERAL OBSERVATIONS.)

ALL hills should be noticed, and their native names discovered and written down. It will be found sometimes that different tribes within view of the same hill will have different names for it.

A rough sketch of different hills and their appearance from different points should be made. Nothing recalls to the memory the look of a hill and the lie of the land so much as a little sketch, however elementary. This sketch should bear the names of the hills above and the place the sketch was made from and the compass direction beneath, like sketch on the following page.

All the most important trees growing in a country should be learnt, as well as their native names. These will be found invaluable, for solitary trees in conspicuous places are used as landmarks. Certain large trees can often be picked out when viewing a

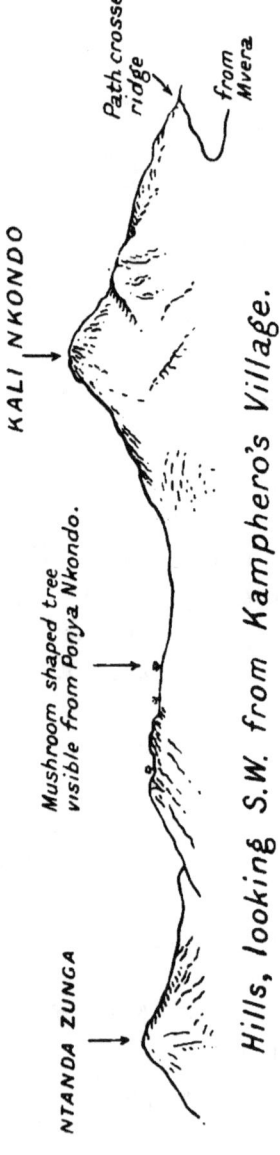

wooded country, marking the neighbourhood of camp or other locality which has to be impressed on the memory.

Groups or belts of certain kinds of trees serve in thick country as landmarks, and are used to round yourself up against, as they leave a good margin for error where a single tree might be missed.

A knowledge of their native names will enable you to point out at once to a native a direction, or to send him to mark the end of a base or a spot with a flag or spear, where it would otherwise be difficult to explain. Moreover, if you see a man, village, or something moving, it is invaluable to be able to indicate its direction without pointing and waving of arms, which might attract attention.

Therefore, we would impress on the reader, on coming to a new country, to learn as soon as possible all the commonest trees and their native names, especially those that grow in belts or clumps, those which grow near water, those which on account of their shape can be recognised at great distances (baobab, palms, acacia), and other trees which grow only in certain districts, on certain soils, or at certain heights above sea-level.

The shape of all important trees marking camps, water, or other localities should be committed to memory, or, if very important, a small sketch should be made of them. Dead trees are often conspicuous landmarks, and can be easily picked out from the surrounding foliage.

It often happens that in thickly wooded country there are slight swells in the ground which give a view of the forest round, and from these places of vantage a tall and characteristic tree marking a known spot can be picked out, though from the spot itself the swell or rise that you are on cannot be seen.

In thick country it will sometimes be necessary to climb a tree to try to get hold of the general features and the rise and fall of the country. Certain trees are characteristic of certain soils, and the soil and the tree together will mark certain localities. Thus we have noticed on the banks of a big African river that where it flows through a certain soil, a tree called " mpani " grows for a distance of about half a day's journey on either bank. In other parts and near other rivers we have noticed other characteristic trees.

Directly one came across this tree one knew that one was in the neighbourhood of the river, and on following the general fall of the country, when baobabs and certain other trees were met with, one knew that the river or one of its tributaries was within a few miles. Beyond the range of the mpani-tree was a belt of thorn bordering it and sometimes pushing into it.

As an example of the use of a belt of thorn, we will suppose that in thick-wooded country you leave camp and march northwards. Shortly after leaving camp you come across a belt of thorn stretching east and west. You continue your march passing into a region of more upland trees. After three hours you have cause to change your course, and make eastwards for another three hours, where you camp.

Next day you wish to return to your former camp. You know from your march of the day before that it must lie to southwest and be distant about four and a half hours. You start off in this direction till you come to the thorn belt in about three and a half hours. You have not yet cut your outgoing tracks, so you know that you must be somewhere eastwards of where you passed through the thorn belt the day before, so you change your course a point or two westward. If after passing through the belt you still have not cut your outgoing tracks, you follow the belt up westwards until you do so.

This sounds simple enough on paper, but in the execution requires considerable judgment and experience, as the thorn belt may be thicker or narrower, and may push into or recede from the upland trees at the point at which you strike it, and all this has to be decided from the lie of the country.

A safe rule to follow in a strange country is, when making for camp, always to allow a large margin for error on the side of your outgoing tracks. If you miss camp on this side, you must strike a tract of country that you have previously traversed, and presumably know, whereas if you miss camp on the other side you pass into an unknown tract and cannot tell when you may be passing it.

Outcrops of rock, lava, etc., may be used in the same way as belts of trees, to round yourself up against, as may also streams and watercourses.

While marching one should always be constructing the main features of the country in one's mind, and where local information is available always ask into what river or lake a stream or watercourse flows.

Where a map is available, by studying it and noticing any track, watercourse, river or path, you may similarly be able to round yourself up against one of these, but maps are often very deceptive, as what may be shown as a big stream may have dried into a watercourse, or vice versa, paths and tracks may have changed and others may have formed.

In hilly country, needles of rock, perched blocks, walls of rock, or bare patches on hillsides should be noticed. When marching on a bearing, whenever a view is obtainable some feature should be selected in advance in the right direction, by which to check your bearing and in case the sun should go in.

Whenever a view is obtainable to the rear, one should carefully study it, with a view to finding one's way back, and should endeavour to recognise all places passed on the march.

When marching on native tracks, they are usually so tortuous that only the general bearing can be noticed. This is obtained by watching the swing of the sun or compass and striking a mean between the greatest variation to each side. All side tracks leaving the one you are on should be noticed, but more especially those forking back from your path, as these are apt to escape notice and on returning the same way will be confusing.

When following a bush path to a distant place, as a general rule it can be taken that where the path forks the less beaten of the two should be taken, as that most beaten leads to a village in the near neighbourhood and is thus more used.

The general arrangement of African paths to distant places is for them to take a wind to pass near a village but not through it, or the village may have been moved to get new ground for cultivation, when paths will be made to and from the old track :

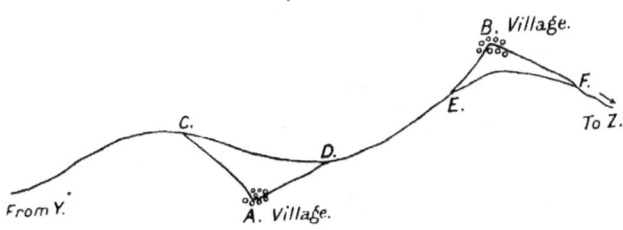

Thus in marching from Y to Z, where A and B are villages, it will be found that the path C A D E B F is much used, while the sections CD and EF are usually hardly worn, and probably overgrown.

At night it may be difficult to keep on a path, as some bush paths are only tracks which cannot be seen at night, and the thick grass may close right across the path. In such a case one will have to go by the feel of the track to the feet, and directly one misses it a halt should be made to find it again. Its windings are usually so tortuous that once it is left for any distance at night it may never be found again. In some places there are well-defined game paths or runs. These generally lead from grazing grounds to water, and may be of help in finding water-holes, etc.

An example of the use these paths may be put to might be instanced. Camp was pitched beside a river-bed with a few pools, but no running water. Along the bank was a well-defined path made by elephant, and on this various landmarks had been selected, among which was at one place a group of three elephant droppings three or four miles from camp.

Returning to camp one night, darkness fell with such black clouds that it was impossible to see any stars or landmarks, and the wind being in gusts from different quarters, there was no possible means of maintaining direction. Following the general fall of the country as much as possible by the feel of descending, the first intimation of having reached the river was one of the natives falling over a bank into it. Having reached the river we did not know whether we were above or below camp, but remembering the elephant path, we felt carefully back from the river with our feet till we struck it. Here, finding some dead grass, we lit it, and searched up and down the path for a landmark, and presently hit on the three droppings, which told us that we were below camp.

Other useful things for marking a track or pathway through thick grass, or where there are no distinctive trees, are gnarled roots or stumps in the path, stones, bones, droppings of leopard, hyaena, civet cat, patches of fungi or ant-hills. White ant-hills are sometimes very tall, and from their summit a view may be obtained, while some are of such peculiar shapes that they make most distinctive landmarks.

In all open places where there are no prominent landmarks close to a place you wish to locate, but distant ones are visible, the place must be fixed if possible after the manner a ship finds a channel or anchorage, viz. by covering of points, preferably two sets of landmarks at about right angles to each other and each set consisting of two marks one behind the other, either covering one another or vertically above one another. The two marks in each set must of course be some distance apart.

If only one mark can be obtained in each direction, the bearing of each must be remembered as well ; while if it is only possible to get one point in all, the bearing and also its distance must be known. As the distance is difficult to judge if the mark is over a mile away, this last method naturally does not fix the locality with any great accuracy. When approaching a camp only marked like this from a flank, don't (till you know the country well) try to steer straight for camp, as when you have come in line with your landmark and have missed camp (as you are sure to do in a country without other landmarks) you will be in the annoying position of not knowing whether to walk towards your landmark or with your back to it. The method of finding the place is to make for a point in the right bearing from the landmark, but at such a distance from it that you are absolutely certain it is either nearer or farther than the place you want; and having arrived at such a position you then walk in the right bearing either away or towards your landmark as the case may be.

Finding the Way in an Unknown Country.

Before starting, notice carefully :

1. The time.
2. The position of the sun.
3. The direction of the wind.
4. Any points near starting-point that it would be possible to recognise again, and mark down position of camp and water, making notes or rough sketches by which to remember them.
5. Any landmarks in the direction of march.
6. General slope and drainage of the country.

While on the march, notice :

1. General directions taken.
2. Any changes of wind.
3. Times taken in traversing distances.
4. Nature of trees and subsoil.

5.   Landmarks of all kinds.
6.   Water or signs of water (especially in a dry country).
7.   Inhabitants or signs of inhabitants (where the inhabitants are hostile it is of course important to see without being seen, and to leave as few traces of yourself as possible).
8.   Places of vantage from which a view can be obtained.
9.   Likely places to contain an ambush or places to avoid, which do not permit of being reconnoitred before passing through.
10. Landmarks to the rear and their aspect, so as to be able to retrace one's steps.

Giving natural features names of one's own very often impresses them on the memory and enables their shape to be recalled the better.

If circumstances permit, a rough field-book should be made on the march. Something after the following style has the advantage of being able to be made without halting. It will naturally be scrawled, but at the first halt a fair copy can be made, and, if necessary, amplified while the memory is still fresh.

The actual times of day can be written down at first, and minus (—) minutes for all halts or time wasted in passing obstacles, going slowly, or leaving the track.

When you make your fair copy, you subtract the minus minutes and show the times by the number of hours and minutes between points.

If a small prismatic compass is carried, bearings can be taken with it from camps and halting-places, while on the march only rough bearings from the sun and swing of the compass can be taken.

Arrows denote fall of ground in the direction in which it points—one arrow gentle, two steep, three very steep fall.

Example of Notes accompanying Field Book.

Breast Hill: Naso Hablod.
Black Rock Hill: Piri Lakuda.
Mtolo camp, half-mile from road.
 Leave road when tall mlombwa-tree covers gap in Breast Hill.
Path to Kalulu leaves road a hundred yards beyond first mkuyu-tree after Mtolo.
Ruzi stream: waist-deep, rocky bottom.
Kalulu's village :
 (1) Needle rock covers peak.
 (2) Breasts as drawn.
Achewa tribal lands commence.

As an example of how to use these notes, we will suppose that the scout has left Kalulu and proceeded eastwards, and then wishes to return

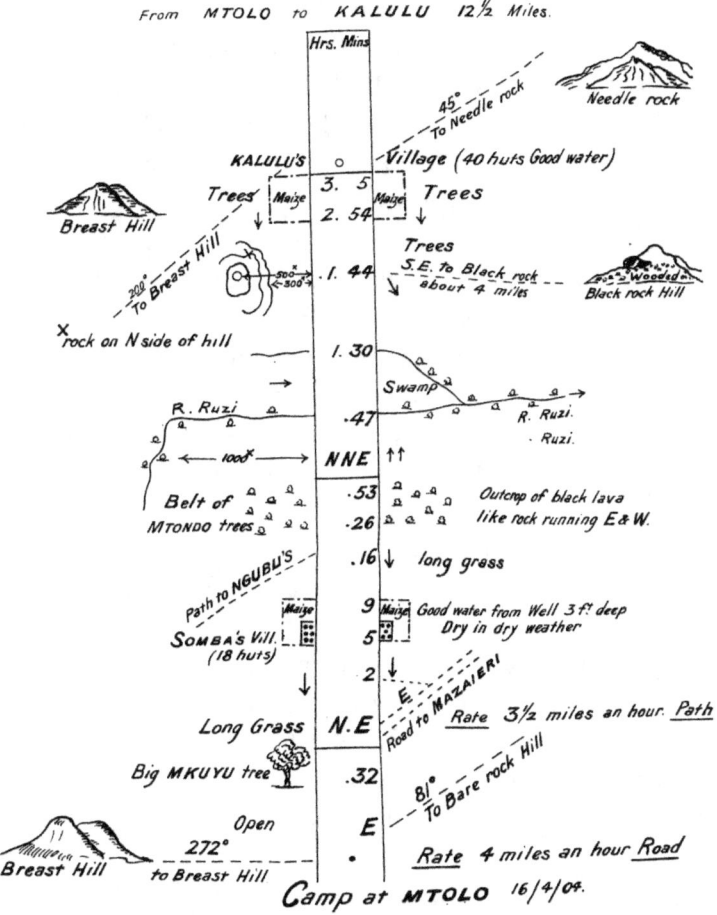

to Mtolo. The country is so thick that no landmarks are visible. He knows, however, that he should strike WSW.

After marching about two and a half hours the

ground rises to his left, but the trees are too thick to see anything. He presumes that this is the lower slope of Black Rock Hill, and that he must leave it to his left. He now knows that he is level with a place on the path about one and three-quarter hours from Mtolo, and that he must meet the Ruzi soon.

After going another hour he still has not met the Ruzi, so he knows that it must have taken a bend southwards. Seeing the ground slope downwards to his right, he turns this way, and soon meets a stream which, by its bulk, he recognises as the Ruzi, for should it be another stream as large, he must have crossed it higher up, either on the march to Kalulu or after leaving that place.

Ascending the other side, he meets the lava-like outcrop of rock and mtondo-trees. Here he would strike WSW., or even W. for safety, so as to strike the path, if he did not know that he must be in the fork between the road to Mazaieri and the path to Kalulu, one of which he must strike, as he has not yet crossed a path or road. Knowing this, he can steer straight for where he imagines Mtolo to be.

Direction can only be maintained by landmarks where the march is from one landmark to another, or where there are a good number, so that positions may be marked by covering of points.

Failing this, bearings have to be resorted to in addition to landmarks. For example, you are at A, and hill B is a landmark. You know that from A you must steer straight past the foot of B, coming into view of your next landmark on reaching E.

Without a knowledge of bearings you would, on arrival at C, follow the foot of the hill round unconsciously and take the line A C D instead of ACE, but by watching the shadows and signs you would not fall into this error.

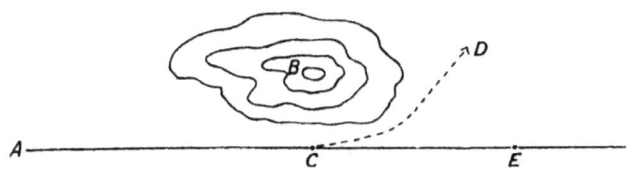

Another common mistake is to overjudge the lateral distance between two distant objects.

Example.—You wish to march from C to a point A, and you know that A is two miles out from a hill B, which you can see.

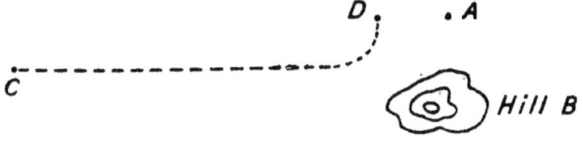

You imagine that at the distance you are from B, A should appear quite close to it (which is not the case), and your march takes the direction of the dotted line on the diagram.

On arriving near B you perceive your error and wheel away from the hill to get your distance of two miles. This makes you approach A from a direction different from that which you expected, and you will probably be looking for A where D is.

In very thick or wooded country what appears to be a well-defined little hill covered with trees may be seen in front, and the trees closing in again, immediately lost to view. On reaching the actual spot, it is easy to walk over the hill (which is only a gentle rise) or an end of it without realising that one has passed this landmark. This slight rise of ground should be carefully looked for, as it affords a valuable landmark in such a country. If when you pass it the ground ascends to the right or the left, this rise might be followed in a difficult country to its highest point, from which it is just possible that a view may be obtained.

In marching on bearings it is always best to work from true north and not from magnetic north. Thus the position of true north should always be kept clearly in mind, and the directions of different bearings, tracks, and landmarks should be noted or remembered with regard to true and not magnetic bearings.

If the scout works chiefly with a compass, he may find it more convenient to note them by magnetic bearings, but whichever he does he should always keep to the same to avoid confusion.

Bearings from sun, moon, and stars will always be true, and to adjust those of the compass he must find the variation. This he can do by taking the bearing of the Pole star, an equatorial or other star rising and setting, or sun rising and setting. The centre of the sun should always be taken.

After a day's march always go over it again in your mind, trying to piece out and remember the look of the country, shape of hills from each point, kind of ground traversed, landmarks, etc., and imagine yourself finding your way over it again and back to the starting-point.

The hardest of all country to find one's way through is the flat wooded country where sometimes a trek of several days is made without seeing a single prominent landmark. For such country an intimate knowledge of the trees of the country is essential, so we would again impress on the reader how important it is to learn these.

A knowledge of trees will also help you to find your way about other less difficult country, and to mark down positions of camp, water, and the enemy. The trees which habitually grow near water are often tall and conspicuous, marking a river bed, and when you notice a scarcity of the trees found in the vicinity of a lake or big river you will generally be right in assuming that you are receding from it.

Sandy desert countries are very difficult to find one's way about in, as tracks are covered up in sand rapidly and there is generally a scarcity of water, but such countries are better than flat thick-wooded country in that there is always a wide range of view, and any few insignificant landmarks there may be are visible for miles, and small hills may be seen a day or two's march distant. In this kind of country also fresh tracks may be picked up and followed easily. There is generally, however, a heat haze which distorts objects, and the eye has to become accustomed to decipher what is seen.

Measuring a Base.

To obtain a base roughly or ascertain the distance between two points, fire a gun at one end and count the seconds between flash and report from the other end. For greater accuracy fire a gun also from the other end and observe from the first end, and then take the mean of the two observations.

" Sound travels at about 1090 ft. per second in calm weather and temp. 32° Fahr. Increases 1.15ft. for each degree above 32°. A moderate breeze accelerates or retards sound by 20 ft. per second " (*Hints to Travellers*, page 55).

To ascertain the time elapsed between flash and report, find out how many ticks your watch makes per minute (150 ticks is the usual rate). With the watch to your ear look out for the flash, counting one-one-one or drumming with the fingers to get the rhythm of the ticks till the flash appears ; then you start counting till the report is heard. The advantage of this method is that the eyes can be kept on the spot whence the flash is expected.

Determining Breadth of Marsh or River.

It will generally be found in tropical countries that the approach to a river or marsh will be covered with dense undergrowth, prohibiting all view. In such a case find the breadth of it by firing a gun as above or by fixing points on either bank as follows : Pace along the approach or track leading across the river or marsh for a certain distance, and from this base fix two climbable trees, one on either side of the path. The distance between these two trees should, if possible, be not less than the probable distance across the marsh. These two trees should then be climbed and from them a point, or if possible two points, fixed on the opposite bank. If the path the other side of the marsh can be seen where it runs into the swamp, this point alone will suffice. The whole can now be plotted out.

In traversing the ordinary native bush- track with compass or plane-table, the grass or undergrowth is often so thick that the direction of the path can only be seen for a yard or two in front, and so forward bearings are impracticable. There may be trees perfectly visible in front, but it will be impossible to tell which will lie on the path or which may be several hundred yards to a flank. In such a case it is best to work only by back-bearings, marching forward till a tree is met with close to the path and committing its shape to memory, march forward (looking back often to see if it is still in sight) till another tree near the path is met, and from here take a back-bearing on to the last tree.

Other expedients are to leave a man holding a bamboo or pole at each station and march on, and when you think that it is about to become invisible take a back-bearing and call the man on to the new station. Bearings may also be taken on to a man on horseback in the same way. To recognise trees for this purpose is easier when one is familiar with the different kinds of trees indigenous to the country.

CHAPTER III.

TRACKING.

> Every scout to be really efficient must be an expert tracker, and this easily comes by practice.—*APPENDIX II.*

TRACKING is the art of reading and interpreting signs or marks left by man or animal after it has passed on elsewhere. As such it consists of:

1. Being able to see, recognise, and tell apart various tracks or spoor.
2. Being able to follow such tracks.
3. Being able to make deductions from them, such as the age of the spoor and the condition, object, and intentions of whatever left such traces.

For the latter an intimate knowledge of the habits and peculiarities of whatever is tracked is required, whether it be man or animal.

Constant practice is required to keep the eye trained to catch the slight marks which to the expert mean so much. Even among savages living in the bush, and gifted as they are with quickness of eye and great powers of observation, expert trackers are few and far between, and almost always come from a family who have been hunters for generations.

The observant man, however, will find that he will make surprising progress in this art, though he cannot expect to equal the native hunter, unless he be prepared to spend many years, or perhaps a lifetime, in the bush.

We cannot pretend to teach a man how to find and follow spoor in a book, but we will endeavour to show him the lines on which to make his observations.

The first thing to do is to be always on the look-out to see spoor. This is by no means easy to do at first, as there are so many other things to look out for at the same time. Till one gets into the way of seeing it mechanically, it is best to always have in mind the condition of the soil that you are passing over, and how spoor would look on it if there were any. This depends largely on whether it is the track of a soft-footed animal (such as a camel or man) or of hoofed animals (such as cattle or goats). The latter are by far the easiest to see, as their tracks cut into the ground.

We will deal in turn with how spoor looks on hard, dry soil ; soft, damp, or muddy soil ; hard soil covered by a thin layer of sand ; soft sand ; dry grass ; and rocky ground.

On hard, dry soil a soft-footed animal leaves very little trace. When the foot meets the ground or leaves it, the foot generally brushes along the surface for a short distance. This may disturb grit or sand on the top, which, on looking closely, will be found to be dislodged ; there may be a little dust kicked forward, a little stone or bit of grit turned over and showing its earthy side uppermost, or what is most common and can be seen quicker by the practised eye is a patch of a slightly different shade of colour than the surrounding ground, when looked at obliquely. This is caused by the turning over of many minute particles of dust or sand, the slight veneer of weather-beaten surface being rubbed off in places and showing the lighter colour underneath.

To follow a track far by such minute signs as these alone is tedious; but there are generally other signs about, which can be picked up by the eye at a distance ahead and followed without loss of time. Such are broken twigs, stray bits of grass or shoots trodden down, earth dislodged from where white ants have covered any object, etc.

With hoofed animals on this ground, besides other signs, the sharp cut of the fore part of the hoof is to be looked for, which is often so clearly cut that it looks like the scratch of a knife. With goats the dragging of the feet is generally very obvious.

In addition to these signs, the droppings of animals can be seen and, with a favourable wind, smelt at some distance ahead, which enables one to cut out the intermediate spooring. On soft or muddy soil the spoor is generally fairly obvious ; but where the ground is only slightly soft, and there is an overhead sun, the spoor of a man is often quite invisible looking downwards, as there are no shadows, but the faint impress can be seen quite easily by looking at the spoor sideways from a short distance off. This is also the case with most tracks on hard soil covered with a slight layer of sand, and such ground should always be looked at obliquely. Sometimes it will happen that, owing to the position of the sun, such a spoor is perfectly visible from one side, though practically invisible from the other.

Spoor in sand can be easily followed, but it is difficult to judge the age. The edges of the spoor fall in, so it is not clean-cut, and the track looks bigger than it really is ; also wind soon obliterates the spoor. Where the grass is dry and thick it will generally be necessary to part the grass at intervals, searching the ground underneath. It can be seen where the grass has been broken down, but great care must be taken not to follow the wrong track, as a great quantity will generally be broken by wind and animals.

The spoor underneath must be seen frequently to be verified. Sometimes the grass will have been pressed into the earth, leaving its impress before springing up again, which shows that something has passed over it, or there may be fresh green grass under the dead, and this may be bruised by anything trampling on the dead grass above. Green grass is easy to follow spoor through, as the grass will be trampled and bruised, and the heads will all be pointing down the spoor.

Where many things have passed through, the track must be constantly verified by parting the grass and looking for the spoor underneath. Both green and dead grass are covered with large drops of water after rain, and in following the spoor at this time it can easily be seen where these drops have been brushed off. With dew on the grass in the early morning exactly the same thing can be observed, and it is then easy to follow the spoor.

The dew or drops of rain will also make the feet of whatever is tracked wet, and so mud, dust, or sand will collect on the feet and be brushed off again on the grass. When the drops have dried, this sand, dust, or mud can be found still adhering to the grass. The same thing is also noticeable in thick bush or among leaves.

It is practically impossible to follow spoor any distance on rocky ground. When it has to be attempted the signs to look for are lichen and small stones dislodged, grass growing in interstices bruised or bent, sand or mud dropped from the foot or body of an animal, or, with man, anything dropped, such as grass, blood from a sore foot, or with tribes who spit much, expectoration and betel juice. Whenever possible it would always be advisable to skirt round rocky ground, endeavouring to pick up the spoor again on the other side.

When the track you are following takes to a hard-beaten track or road, the quickest way is to follow the road, searching only every turning or possible exit to see if the road has been left. On long strips without exits the spoor should be verified every now and then by searching carefully for it, to make sure that it has not left the road unperceived.

Determining Age of Spoor.

If the ground is hard and caked, and spoor is seen of the soft-ground type, it is obvious that the ground must have dried since the spoor was made, and the time can be approximately judged. If there has been rain lately and yet no drop-marks are seen on the cut or pressed portions of the spoor, it is likewise obvious that it must have been made since the rain. If there is sand or mud sticking to the grass, as described above, the man or animal must have passed when rain or dew was still on the grass.

Spoor loses freshness and its clean-cut appearance with age. This varies for different conditions of soil, sand, wind, sun, shadow, or rain, and so must be learnt by experience. Only in very barren country can anything go far without treading down a blade of grass, flower, weed, or shoot. These should be looked for and carefully observed when found, as the sap at the bruise or break gives the necessary indication of age.

Hoofed animals often cut bits of grass, etc., clean off when treading on them, and animals while grazing drop leaves or blades of grass on the ground. These often show the froth from the animal's mouth, or, failing this, the sap at the break can be observed. How much this moisture has dried or how much the leaf or blade has withered is a very sure guide to the age of a recent spoor. The amount it has withered depends on the heat of the day, whether it is dropped in shade or sun, and the kind of plant.

Men whilst walking along often pluck or break twigs, shoots, and grass—sometimes to keep them off the face and sometimes merely from habit.

Droppings when found are the surest and easiest means of determining the age of spoor, and will be treated later. Whenever tracks pass over favourable ground, where the impress can be clearly seen, their shape and peculiarities should be learnt by heart. The different points to notice in different spoors are :

Man *(bare-footed)*.—Length of stride, length of foot, breadth across ball and heel, relative length of toes, the arching of the instep (i.e. how little or much shows between the ball of the great toe and the heel), the shape of the ball, the angle at which the feet turn out from each other, any cracks in the thick skin of the heel.

If the track is one you especially wish to remember, the measurements should be written down, not forgetting to add the sort of ground that they are measured in. Naturally a spoor on soft mud or sand is bigger than that on hard soil.

Negroes and negroids are very wide across the ball of the foot, and this part is very prominent and strongly curved. A woman's foot can often be distinguished from that of a boy or young man, as it is not so broad in this part.

The Arab has a small and well-arched narrow foot, while the woman has often a straighter line on the outside of the foot.

The Somali walks more flat-footed than the Arab, and has a larger foot.

Indians, as a rule, have narrower feet than Africans, but are wider across the toes than Arabs.

Negroes have much bigger feet than negroids.

The footgear of different nations should be studied, such as the goat-skin sandals of the Bedouin with flaps to frighten away snakes, the leather madaas of the townsman, chaplis of the Panjab, grass sandals, wooden sandals, pointed leather shoes, etc.

Man *(with boots or shoes).*—Length of stride, angle at which toes turn in, side on which the impression shows deepest; these remain fairly constant whatever the footgear worn. Also observe length of boot, breadth of heel and toe, pointedness of toes, soles nailed or sewn, arrangement of nails in the heel, heel and toe-caps, non-slipping pads, etc. When running, the stride is longer and, as a rule, only the toes show, and with bare feet the toes splay more.

Cattle.—Cloven-hoofed, clean-cut spoor, very round in front. Learn the difference in size between bull, cow, and calf. Native cattle generally much smaller than European or South African. Learn the relative sizes of a full-grown bull of different kinds. It may prove useful to know what kind of cattle have left their spoor—for instance, whether certain tracks are of cattle which have been looted from a white settler. Following up cattle would, of course, lead to water and their kraal, which might be a stockaded village.

As flocks and herds are the sole wealth of many tribes, an expedition against them, when the enemy is more mobile than the attacker, generally resolves itself into capturing his stock. As its loss will be severely felt by the native, this will generally have the effect of causing him to make an effort to recapture it, when a decisive engagement may be fought. Thus an important part of any such expedition will consist of scouting to find flocks and herds or their tracks, which are then followed up.

Tracks of animals can be followed by browsing as well as by spoor-marks. The difference between cattle and horses browsing can sometimes be seen by the marks of their different dentition, cattle having no front teeth in the upper jaw. Country grazed over by donkeys will be cropped closer to the ground.

Cattle on trek from one locality to another will not leave the closely trodden track of the herd, while during grazing they will be seen to open out. If the grazing-ground is only of limited area it may save time to cast round the edge of it, in the most likely direction to find the tracks of the herd on leaving it.

Sheep. — Pointed, fairly regular spoor ; they walk slowly, and only cover short marches during the day, so will always be close at hand when fresh spoor is found.

Goat.—Untidy, irregular spoor. Being by nature a mountain animal, with feet adapted to rocky country, the hoofs grow faster than they can be worn down when they are kept in lowland countries, and where there are no rocks. This causes them to assume irregular shapes, and they frequently have long, twisted, and distorted hoofs. When on the march they drag their feet, scratching long lines on the ground.

Horses.—May be shod or unshod.

At a walk the tracks of the near feet are together and those of the off together.

At a trot the stride is longer, and the track of the hind is just behind or over-riding that of the fore.

At a gallop tracks much in the same line with one in advance. Earth much scattered and thrown to the rear.

Tracks of individual animals are told in the case of unshod animals by the splittings of the hoof and by the depth and shape of the impress of the frog, which varies enormously. Tracks of shod animals are very difficult to tell apart, as so many shoes are made after the same pattern, and the frog is constantly being cut. The position of the nails, the length of stride, whether hoofs of hind turn inwards before being raised from the ground, and the lateral splay of the feet are used to distinguish different animals.

It must be remembered that the lateral splay differs according to pace ; as a general rule, the faster the pace the nearer all four feet come into one line.

Camel.—Round, soft-footed spoor, with two toe-nails in front and an inverted V-mark behind. This V is generally cleaner cut and sharper with a trotting and well-bred camel, while with baggage and other camels it is more indistinct and sometimes almost a semi-circle.

Learn the difference in size between male, female, and young. On hard ground sometimes only the marks of the toes are visible, and then appear as two dots side by side.

The bottom of the foot is covered with hard skin, which peels off at intervals, leaving little raised surfaces or depressions. It is by the peeling of the foot that spoor of individual animals is easily and unmistakably told. The impressions and the relative positions of these can be seen in the spoor, and the feet on which they occur should be noted.

When many camels have to be recognised, a small drawing may be made of the mark and the foot, writing underneath on which foot. As these marks are always gradually changing, it is necessary to keep them up to date.

If you were very anxious to recognise a certain camel's track easily, it would be possible with a penknife to flake away a little bit of hard skin on some part of one foot, as also with a horse to pare away a small part of the frog or hoof.

Donkey.—About the size and shape of the heel plate of a boot reversed.

Pig.—The two sides of the cloven hoof are

each semi-circles, and so the spoor is round in front, and not pointed or angular like a sheep or goat. As it is often the case that only certain tribes keep pigs, it might be useful in a deserted village to recognise their spoor.

Dog.—Learn the usual track of a pi or pariah, and compare the difference with that of civilised breeds.

The usual difference between the spoor of fore and hind feet in most animals is that the hind is thinner and more oblong, while the fore is rounder and thicker. With cloven-footed animals the inside half of the hoof is generally the shortest, which may enable you to tell whether a single spoor-mark belongs to near or off foot ; but this rule is not invariable, as occasionally the outside half of the hoof is the longest.

There are certain wild animals whose tracks may be confused with domestic animals. Such are buffalo, larger than native cattle spoor. Eland female and young more pointed than cattle, but the old bull very like. Zebra almost exactly like that of a pony. Zebra are generally found in herds, and often take to the hills during the day. Ponies grazing in numbers would have men with them, whose tracks would be seen, and they might be hobbled. Wild donkey exactly like that of unshod domestic. Pig like that of boar, bush-pig, and wart-hog, but the former would always be found in or near villages. Sheep and goats something like certain buck who go in herds, but as a rule less graceful and slender. Dogs like jackal, but latter are narrower, and almost always have traces of berries in their droppings. Hyaenas bigger than village dog, but very like boarhound, but the former have white droppings.

The length of stride of different animals at different paces should be known roughly, as then one knows how far in front of each spoor to look for the next. Some native trackers follow spoor by pacing the length of its stride, putting the foot behind each spoormark alternately.

Where there are cattle and sheep there is presumably water quite close, as they are accustomed to drink daily. It can be seen if they have drunk recently by how muddy the water is, and how much sediment is mixed up with it.

The vehicles of the country should be studied. Points to notice : if four, three, or two-wheeled, or sleighs, what kind of draught animal, breadth between wheels, etc. Rate of going can be obtained from the pace of animal.

Other Signs.—Axe-marks on trees. To practise telling the age of these and also of the broken twigs or shoots trodden down, make marks or break off shoots at different intervals of time, and then compare those of one, two, and three days old, and also those of a few hours and fresh ones.

Froth from horses or cattle, showing latter to be hard driven.

Expectoration of men, smouldering embers, blood spoor, staling, specks of mud, from all of which the age of the spoor can be judged:

In addition to these are encampments, bivouacs, and the numberless small things found in deserted villages or halting-places, and even on trek, such as grains of rice (cooked and uncooked), beans, flour filtered out from the bottom of baskets, and beer or water spilt out of jars while being carried, feathers from head-dress, beads, oil, buttons, rags, tobacco, berries plucked by the way, etc.

Droppings. Every man and beast is forced to leave these traces of itself at intervals, and seeing whether they are still warm, soft, hard, or only hard on the outside, gives the surest indication of the age of the spoor. They also show the condition of whatever deposited it, and how it has fed. Different nationalities perform this office in different ways: Arabs use sand, Indians water, Boers stones, and Africans generally omit this portion of their toilet altogether.

Traces of jowari or grain will be seen in the droppings of grain-fed camels, by which means the spoor of any of the camel corps could always be told from a Somali baggage camel, the latter being fed entirely by browsing

As an example of how the customs of a people must be known before much useful information can be deduced from spoors, we might instance the case of a donkey's spoor being found, and at once attributed to a wild donkey. The reason for this assumption was that it was not the custom of the country for men to ride donkeys, but they were used for baggage purposes, and were occasionally ridden by women. In either of the latter cases there would have been a person on foot accompanying the donkey, whose footmarks would have been visible.

CHAPTER IV.

GENERAL HINTS.

FIRST, as to clothing, the scout should naturally be as inconspicuous as possible, and adapt his clothes to the country he is in.

Whatever he adopts, it is the greatest mistake to be clad in one uniform colour, however suitable it may appear, for in certain lights certain colours show up more than others. Whenever he is in a light which shows up his colour, his figure, as a whole, stands out; whereas were he wearing different shades, a patch showing up distinctly would not give to the observer the idea of a man's figure.

Nothing at first sight would appear more conspicuous than a piebald horse or a leopard, yet when standing still in deep shade, and especially at dusk, these animals are far less conspicuous than a uniform-coloured animal, as the eye catches either the lighter or the darker blotches or patches. The darker parts of the piebald might be bushes and the lighter the spaces between them, and the two have to be connected before the outline of a horse is detected. If a black horse was mistaken for a bush, the first impression would be how like an animal that bush was. White is, of course, the worst colour possible ; but if the white parts of the piebald were dirty green, this animal would be quite invisible at a few hundred yards when standing still in bush country. All animals and birds who rely on their colour for protection are lighter coloured underneath. The reason for this is that the underneath parts are more in shadow, and hence must be a lighter colour than they are intended to look. So with the scout it would be as well to have his breeches of a lighter colour than his coat.

The best kind of dress is :

Hat—brown soft felt, with brown or khaki puggaree for tropics. As it will probably be used as a pillow, it will soon get discoloured with earth if too light.

Shirt—dark brown, with breast pockets and a few loops to put cartridges in if required.

Coat—mottled greenish and brownish, without any distinct hue, and should be well discoloured with age. Two leather shoulder-pads will further help to break up the outline. It should have plenty of pockets, both outside and in ; some at least should be lined with waterproof to keep matches and note-books dry, and these should be protected with flaps. Have pockets and several ticket pockets for compass, watch, etc. Leather knob or button on shoulder, to prevent rifle-sling slipping off, should turn up well round the neck when required, with a flap or strap to button across the throat.

Scarf—khaki silk.

Shorts or breeches—Khaki cut-shorts for walking or breeches for riding. These should have hip pockets for revolver or note-book.

Belt—strong leather, with small pouch and sheath for knife.

Boots or shoes—light, with thick rubber soles.

Putties—khaki-coloured.

Riding or field boots should not be worn, as one cannot walk far or run in them.

The scout should have with him :

1. Rifle and ammunition.
2. Compass.
3. Watch.
4. Box of matches.
5. Strong knife.
6. Note-book and pencil.
7. Emergency ration.
8. Field-glass.
9. A native guide or tracker, who will carry—
10. Water.
11. Canteen.
12. Axe.

13.   Food ration or strip of biltong.

1.   **Rifle**—The rifle would only have to be used at close ranges in case of surprise or other mischance, so a sporting rifle or carbine would be handy, good for rapid shooting, and light to carry. It should be provided with a sling. This would usually be sighted up to 300 yards. It could have additional sights put on if thought necessary; but it must be remembered that a man cannot be considered a certain shot over 150 yards, even to an expert game shot, and the scout would not want to give away his position, waste his ammunition, and run the risk of the man shot at getting away and giving the alarm, so would seldom shoot at over this distance. In ordinary country at this distance, for a man facing you, aim at the middle of the chest, so as to penetrate heart or lungs. For a side shot aim underneath the arm-pit. For longer shots in thick or bush country, where the common error is to over-judge distance, aim just above the navel ; and in clear, rare atmosphere, where under-judging is common, aim at the base of the neck. To finish off a dying or wounded animal, aim from behind at the back of the head, where it joins the neck, so as to rake forward into the brain, or, missing that, cut the spine. If the ammunition is in clips, they should be constantly looked at to see that they are not blocked with rust or dirt, as this may cause a delay or jam in putting in. The rifle should have a non-automatic safety-bolt or catch, and be carried loaded with the safety-bolt on. At night the rifle should be laid close to the body, on the right side. If a revolver is used, the best place to put it at night is under the right thigh, attached to the wrist by a lanyard. In the case of the ordinary Service revolver, it should be loaded with only five cartridges. A Colt's automatic pistol will be found handier, as it will lie flat against the side.

2. **Compass**—A small pocket prismatic, about the size of a large watch, is the best. It should have points of luminous paint, and can be used during the march as an ordinary- compass and at halts as a prismatic. The new Service compass (Mark V.) is a good pattern (see *Manual of Map Reading and Field Sketching*, 1906, page 39).

3. **Watch**—Watches are constantly going wrong in damp or tropical countries. A half-chronometer is used for astronomical observations and can be relied on, but the price is rather prohibitive.

4. **Matches**—should be in watertight box or waterproof envelope.

5. **Knife**—Strong shikari or cooking-knife, in sheath, is as serviceable as anything.

6. **Note-book**—Thick waterproof cover, with flap. Part ruled in squares to convenient scale, and also blank sheets and pockets. Spare pencil.

7. **Emergency ration**—Should be portable and compact. Depends on the country. Meat and bread or biscuits, or 2 lb. of dates, are most sustaining.

8. **Field-glasses**—should be prism

glasses Finding objects should be constantly practised as field-glasses cannot be used to their full advantage otherwise. If the scout knows how to use a telescope he will probably prefer the latter. A telescope in skilled hands gives incomparably better results than field-glasses, but has the disadvantages of being less handy and steady. It is very difficult to use a telescope of great magnifying power in a wind or haze, and moreover much practice is required before it can be used quickly and thoroughly.

9. Native tracker—Should be strong and reliable, and a hunter by profession if possible. If he is accustomed to use a bow he should bring one with him, as with it food may be got where it would be inadvisable to fire a rifle.

10. Water—in big aluminium water-bottle, the size depending on the country.

11. Canteen—Infantry canteen as useful as any. Should contain packed away inside small tin of cocoa, saccharine, salt, pepper, small bottle of sauce (if room), small tin of barley, quinine, spare box of matches, and any spaces filled in with biscuit.

12. Axe—Should be of native make, as the native will be able to use it better than one of European make.

13. Food ration—for native ; rice, flour, or dates for himself and a strip of biltong for both. Biltong is made by boiling rock-salt in water. Skim the scum off top. Cut the meat up into thin strips, tie a bit of string on to the ends, dip them into the boiling brine for a few seconds, pepper well, and hang up in the shade of a tree for two days.

When mounted, there should be about five yards of rope wound loosely round the animal's neck, to hobble him with. A blanket may be worn under the saddle in place of a numnah, and can be used by the scout at night. In barren country several pounds of grain should be taken for the animal.

Other things which should be taken are : A little potassium permanganate in pouch of belt for washing wounds or in case of snakebite, field-dressing, magnifying glass for lighting a fire or pipe, or reading native writing (this might be unscrewed from field-glass, but would be liable to get lost), a few yards of strong twine, pipe, tobacco, flannelette for rifle, oil-bottle and pull-through in stock butt trap.

A rough elementary idea of geology is useful, especially as regards the clays which throw off water and the sandstones, chalks, etc., which hold it, so as to be able to locate water. Pools may be expected in low-lying places in clayey soils, and water may be obtained by digging in a water-holding soil known to have a stratum of clay underneath.

If a village is noticed near a dry river-bed and the inhabitants are hostile, digging in a sandy bottom of the bed below the village at night might be productive of water. Other indications of water are doves, always close to water (called in Somali " dweller at the wells " for this reason), certain buck which are always found near water, game tracks which lead to and from water, certain reeds, and during the dry weather, when all the grass is burnt, a patch of green grass often marks the position of a tiny hole. Such a patch should be examined closely, as it may conceal a little hole covered over with vegetation. Certain trees which only grow near water.

Certain waters leave deposits where they have been lying, and very often these deposits are white and can be seen from afar. From this, places where water is lying or is likely to be found lying may be told. If there is no water on the surface there may be some just below it. A hole should be scraped out, and if there is any moisture, it may be lined with reeds, to filter the water when it comes and keep the mud down. Sometimes muddy pools are found so shallow that the sediment is at once disturbed on trying to drink or draw off water. In such a case a floor of grass or reeds may be made for the water to filter through and to keep down the mud, so that comparatively clean water may be drawn off above, or it may be sucked up through a straw.

For a bivouac a few branches or saplings may be cut down and tied with bark rope across other trees standing close together. Then tie long grass, branches, or reeds on to this framework so as to make a shelter to windward. The trees which give the best bast or inner bark for this and other purposes will always be known to the natives, and should be noted, as also the trees from which string or fibre can be made.

If a fire is lit at night, it should be in a screened place such as forest, tall grass, or cave, or it should be covered round with branches so as not to be visible. In the season of bush fires a fire will naturally attract less attention. In the daytime it should be made generally among trees. Smoke does not show much under trees, and is scarcely perceptible under certain kinds of whitish thorn, which themselves look like smoke at a little distance.

The bivouac, however, should be near a position from which a view can be obtained to prevent surprise. If no such place can be found, perhaps a camp among dead grass, reeds, or cane might be advisable, as any one approaching could be heard a long way off when you are sitting still, whereas they, from the noise they will be making, will be less likely to hear you when you move off. The grass might also be fired to cover your departure. To do this, have a bundle of dry grass ready, and putting the end in the fire, run along across wind, dropping small burning pieces at intervals, wherever there is a likely tuft.

Care should be always taken to leave as few traces of yourself as possible, more especially before camping or halting. Where you cannot help leaving a conspicuous spoor, it would be as well before halting to make a detour so as to come back within sight or earshot of some place that you have already passed. Any one following on your spoor can then be detected and eluded.

If the night is chilly, the putties may be taken off and wrapped round the stomach and neck to avoid getting a chill.

Any fresh meat obtained can be cooked by

placing it in a cleft stick, the end being sharpened and stuck in the ground near the fire. This can be turned at intervals. Soup may be made from the biltong, barley, and sauce. When the fire is once started, three biggish saplings should be got and placed with their butts in the fire, in such a way that, as the ends are consumed, they can be pushed farther into the fire and kept all night. A little pile of sticks or dead wood should be kept handy so as to restore the fire when it shows signs of going out. Both these and the fire should be within reach of where one lies, as it is fatiguing to get up often during the night to feed the fire. It must be remembered, however, that the sound of an axe cutting wood can be heard an immense distance.

The horse should be hobbled and allowed to graze during any halt in the day and the first part of the night. Unless the horse is much in want of food, it would be as well to bring it in and tether it before sleeping. Which way the horse is grazing should always be noticed, so as in case of alarm to know which way to run for him. In such a case, if hard pressed, pull out your knife as you run to cut the hobbling ropes. Till fresh rope can be obtained, a bit of bark rope will then have to be twisted up.

As lack of vegetables or green food produces veldt-sores, scurvy, and other undesirable complaints, all indigenous vegetables, and especially wild ones, fruits, and berries should be inquired into and their native names made a note of. In many countries there is a wild spinach which is welcome when no other green food is obtainable.

Long sight, quickness of perception and observancy are all things of great importance to the scout, and can all more or less be improved by constant practice. The art of picking up objects at long distances is almost entirely a matter of habit, as is also the picking up of objects with field-glasses, and to use a telescope with any effect needs constant practice.

How to quickly and accurately describe the position of a place must be learnt. Directions are often given in the most ambiguous way. If some object is seen in the distance, to rapidly and clearly explain its position is a thing that requires repeated practice. The best method is to first point out some tree, bush, grass, or stone quite near to give the true line, then state the number of rises and falls of the ground to it, and then the approximate position on the rise or fall it occupies and something near it viz. " Straight over that umbrella-shaped thorn-tree, about half-way up the third rise, there is a bushy tree. I saw something moving to the left of that." In flat country it is necessary, instead of the number of rises, to give the approximate distance or point out something the same distance away.

In a dry country only drink from your water-bottle under urgent circumstances. When good water is met with, refill, or at least damp the felt cover of the water-bottle to cool it. If water is met with in a hot country an hour or two before arriving in camp, it would be as well to drink, as otherwise, arriving in camp with a tremendous thirst, you would be unable to eat.

The ability to go long days without water and food is largely a matter of habit and custom. Most people indulge in three heavy meals a day, and this develops into such a habit that when they have to go without, the stomach demands what it has been accustomed to, and so much inconvenience is suffered.

If one accustoms oneself to miss meals, there is no violent demand of the stomach, and hence no inconvenience. Those who have never practised this should not begin suddenly in a tropical climate, as fasting in a hot sun is apt to produce fevers, sunstroke, etc.; so the practice should be gradual.

Sleep also is a matter of habit, and can be curtailed or taken in little pieces, or on the back of a horse or camel, when one is used to it.

If much work has to be done on foot, boots and shoes should be very light, and, needless to say, comfortable. As silence is a great feature in scouting work, they should have rubber soles. If liable to suffer from sore feet, they should be soaked after marching in salt water or solution of permanganate of potassium, and before starting for a march the feet and socks should be smeared with grease, fat, pomade, or vaseline. The latter is also useful if one is liable to chafe. The boots should always be greased as soon as possible after getting them wet. Any fat, such as fat-tailed sheep, hippo fat, etc., can be used, and only requires boiling down. Beeswax is also useful, and a mixture of fat and beeswax boiled together is excellent.

To get boots wet and allow them to dry on the feet while marching spoils boots and produces sore feet, so one should endeavour, if possible, to keep boots dry, even if one has to take them off to cross streams; and if it is unavoidable to get them wet, such as walking in dew, they should be greased at the earliest opportunity. Constant changes of boots relieve tired and sore feet. A light pair of shoes might be carried as a change on a long march. If the instep is swollen, it may give relief to fasten the boot outside the putty for a while. Boots should be laced tightly at the bottom and loosely round the ankle.

When on the march the eyes should always search in every direction for any signs of movement or traces of human beings, and the ground in front and at the sides should be examined for spoor. Where spoor is of importance avoid rocky country and endeavour to pass over ground on which it can be easily seen. Always remember to keep out of sight as much as possible, and not to make a noise by going through dead grass and reeds or treading on sticks or dead leaves.

Where the native of the country wears any kind of shoe, sandal, or footgear, the scout should be provided with a pair, and if they are too uncomfortable to wear always, he should carry them with him to wear whenever it is of importance that he should not leave his own tracks. Needless to say, the change should be effected if possible after traversing rocky or hard ground. When approaching open country, one should go very slowly and cautiously, not committing oneself to the open till every direction has been scrutinised through glasses. If possible, it is best to skirt round cover.

A private code of signals and signs should be arranged with the native tracker or between scouts accustomed to work together, so that one can communicate at a distance or without speaking, and also that if separated one may be able to follow the other. These signs should be quite different from the usual bushmen's signs of the country, or else they run the risk of being interpreted and possibly altered by the enemy. The usual signs should be ascertained, so that they can be read when seen, and if necessary changed.

When scouting in difficult country, the scout should always have his line of retreat mapped out in his head for every possible contingency, so that if he suddenly comes on the enemy he will be prepared to act without hesitation, If a new vista of view is opening out he should move slowly in that direction, stopping frequently, remembering that a man standing still, especially in shadow, is most inconspicuous, but if moving can be picked up by the eye quickly. The next best thing to standing still is to move straight away from or towards the observer, being yourself in shadow or covered by a dark background, so move slowly straight towards any opening view, to be less conspicuous to any one who may be in front.

If you wished to make a mounted man a prisoner it might be necessary to shoot his horse under him, for which a raking shot from either front or rear is best. If it is necessary to shoot from broadside on, the best place is behind the shoulder. Following on blood spoor, frothy blood indicates a hit in the lungs or windpipe.

Breaking twigs and blazing trees sometimes have to be resorted to, so as to mark places or routes. Thus, passing through dense undergrowth or wood without sun to direct you, it may be advisable at intervals to half break a branch, leaving it hanging, so as to mark the way back. A branch hanging like this, with the leaves reversed, is a sign which can be picked up quickly by the practised eye.

CHAPTER V.

TRIBAL CUSTOMS AND DIFFERENCES.

For scouting and reconnaissance work no pains should be spared to learn the languages, customs, habits, temperament, character, and peculiarities of the people or tribes you have to deal with. Till this is done it will be impossible to make correct deductions from their signs or tracks or to gather much valuable information.

A few examples may show how such knowledge may be turned to account:

(1) An earthenware cooking-pot found in a deserted karia or camping-ground. In Somaliland the inference would be that the departure was made in the greatest haste— perhaps in panic—as the grasping Somali would never part with anything, however valueless, if he could possibly help it, and the idea that anybody might find it and get it for nothing would be most repugnant to him. With a careless, good-natured, happy-go-lucky people such as the Burmese or the

Ayao of Central Africa a similar find need bear no such interpretation.

In Somaliland, on returning to our zeriba, the track of a horse walking was found about four miles from the zeriba. The inference was that it was an enemy's scout, as the usual pace is a jog. Moreover, had it been one of our irregular scouts, he would have broken into a gallop on getting so near camp, to get in the sooner, whatever the state of his horse, as the Somali is callous to the sufferings of his animals. The track was quite fresh, and, on being followed up, the inference was found to be correct.

A bit of crocodile foot, in a deserted village near Lake Bangweolo, proved that it belonged to the Awisa tribe, as, though resembling another tribe in most things, they differ in that they are the only tribe in the neighbourhood who eat crocodile meat.

A man seen on a camel half a mile or more away in South Arabia was recognised as being of slave blood, as he was swinging his legs. Had he been an Arab he would have kept them close to the animal's side.

The customs of war of the tribes to be encountered are naturally important, but other customs are no less important, as many deductions may be made from them. For instance, from the customs and habits peculiar to different tribes it might be ascertained how many tribes were implicated in or in sympathy with any native rising, the extent of which was not as yet fully known.

The customs connected with smoking differ very widely. In Africa some tribes prepare their tobacco leaving in the ribs of the leaves; some take these out, some roll the leaf into a ball, others into a stick tied up with bark, some take snuff, others chew, some mix their snuff with ashes, others with powdered snail-shells, and others again do not use tobacco at all.

In Asia some smoke the pure leaf in the hookah, while others mix the tobacco with molasses or with incense ; some use pipes, some use opium, and others make cheroots or roll chopped tobacco in leaves. Some smoke and do not drink, as is the case with most sects of Mohammedans ; one sect neither smokes nor drinks. Some drink but do not smoke—eg. the Sikhs. Some eat kat when smoking.

Nearly all castes or tribes have some distinctive caste or tribal mark, and this often varies with the sexes. Such marks are made with indigo, henna, and various other substances, or by tattooing or merely by making scars, lumps, etc.

In matters of food, tribes, and often even subsections of tribes, differ. These differences are generally connected with religion. Thus in Asia one finds large numbers of people of each religion who are under the same laws as regards food—e.g. Mohammedans, Hindoos, Jews, and Buddhists—but for each of these religions it will be found that among different tribes and peoples there are different interpretations of these laws. In Africa all the tribes who have not come under Mohammedan influence, and even some of those who have, possess some fetish animal which a tribe, or some part of a tribe, may not eat. These may be almost any animal from an elephant downwards. These animals are in addition to any animals which from sanitary or other reasons they may not think fit to eat.

Thus a tribe called Awemba may not eat crocodile, as it is their fetish animal ; a section of the Achikunda tribe may not eat baboon for the same reason, but do not eat crocodile also, as they think it unfit to eat.

Weapons differ with different tribes both in form and in pattern. Some use only bows, some only spears, and some both ; some use shields and some do not. The arrows may be feathered or unfeathered, and perhaps only made of a certain kind of wood. The spears may be short throwing, long stabbing, or both carried together. Feathered darts may be carried in addition. Different tribes may have different poisons for their arrows. Shields vary very much in form, and sometimes different sections or families of the same tribe are distinguished by their shields bearing different devices—e.g. the Masai. Rifles may be of one pattern, as the Arab matchlock, but are more usually of any European pattern obtainable. The powder-horn differs both in make and in the method of being carried, some wearing them in front and others at the side.

The Berbers are said to have different devices on the rear of their goat-skin cloaks to denote different sections. The Abyssinians have bands of silver on their sword-scabbards to denote the number of men they have killed. The Masai bear different devices on their shields to denote different families.

Almost all tribes, however savage, have a certain amount of knowledge of handicrafts, such as rough iron-smelting, making salt, mats of different designs, bark-cloth, clay pots, etc. Some tribes are dependent on others for certain of these luxuries. It often happens that a dominant fighting tribe is dependent on members of a weaker tribe, living among them, for things of such vital importance to them as spears and arrowheads. Certain sects of Mohammedans do not keep dogs in their villages.

Different peoples have different methods of making fire. Many still use two fire- sticks ; others flint, or snap their flint-locks over a scrap of cotton waste or fluff, torn from the turban or loin-cloth. Some peoples chew betel or areca, the red juice of which can be seen on the ground where it has been spat out. Articles of toilet or furniture may be found in an old encampment, such as toothsticks, combs, beads, wooden pillows, spoons, etc. Grains of rice, beans, the white flour of maize or the red of millet may be found on the road, the latter showing that there were probably women with or near the party, as in Africa men do not pound grain.

Different signs used in the bush, such as blazing trees, smoke signals, etc., should be learnt and looked for. Many African tribes have a way of showing followers or stragglers which way the party in advance has gone by " closing " all paths not to be taken. This is done by plucking green grass, leaves, or twigs and throwing a few in the mouth of each turning not to be taken, or by scratching a line across the path with a spear or stick. Freshly-picked shoots or grass are always used for this purpose, to show those behind how recently they have passed, and also so as not to be confused with any previous signs there may be.

When the path is to be left for the open, the path is closed as before, but this time generally with a big branch, so as to be observed, for a few leaves might escape notice elsewhere than in the turnings where they would be looked for. This branch shows where to leave the path. After proceeding a few hundred yards a branch of a tree is broken and left hanging, or a knot is made in a tuft of grass, to show which direction to strike across country. Care is taken that these signs are visible from the path. If after proceeding across country another path is met with, if it is to be taken a hanging branch will be left beside it, but if not to be taken it will be closed above and below where crossed.

Ghurkas have a method of showing which path to take at night, when the road forks, by twining a twig stripped of bark and leaves into the undergrowth at the side of the path a few yards from the fork, and pointing in the direction to be taken. This can be felt for in the dark.

It will be found that different tribes living in sight of some great natural feature often have different names for it, while those living a little farther off have a very hazy idea of it, and will often give the name of some big chief living in that direction as the name of a hill or river. Native names should, therefore, be written down, with a description of the feature and the tribe of the man from whom the name was obtained.

Natives living on the shores of a great lake or river as a rule call it the " lake," " river," or " water," as it is the only one known to them, just as a man living on the coast of England says that he is living by the sea, instead of calling it the Channel, Atlantic, or North Sea. Thus the words " nyanza," " nyasa," " mweru," " dweru," and many others merely mean " lake " or " water " in different languages ; whereas the same river often has a different name in each country it passes through—eg. Chambezi, Luapula, Lualaba, which are all really the River Congo.

A native's instinct can generally be relied on to take him back to a place he has already been to during the daytime, and especially if he be a hunter, but at night they are notoriously unreliable. If still on the march at sunset, the guide should be made to point out the direction with a spear, and the bearing of this direction should be taken by compass, sun, or stars. Having done this, when darkness has set in it would be as well to ignore whatever the native says, however well he says that he knows the way.

There is generally a great difference between a hillsman and a plainsman in describing or pointing out the way to a distant place. The plainsman, being accustomed to march for days by the sun and bearings alone, will point out the direction to any place that he may know well in an almost perfect compass-bearing, while the hillsman will brandish his arms all round the horizon, but if what he is saying meanwhile be carefully noted, it will be found that he is giving a very clear account of the landmarks to be looked for and how they are to be used.

CHAPTER VI.

RECONNOITRING HOSTILE KRAALS OR VILLAGES.

Reconnoitring is done by day chiefly by observing from a distance and noting tracks, and at night by listening and smelling. The presence of man in the neighbourhood can be detected by:

Seeing his spoor.

Hearing him, his cattle or dogs. An axe can be heard easily two miles away on a still day, and drums more than twice that distance at night. Natives, moreover, are usually very garrulous.

Smelling his fires or stock.

Seeing smoke.

Birds or animals being frightened, such as spoor of galloping game, pheasants, partridge or guinea-fowl being disturbed, plovers and peewits calling, especially during the nesting season, honey-birds twittering, birds flying up from trees, such as rooks, vultures, etc., and in various other ways.

The presence of the scout, in the same manner, may be given away by such animals or birds, and so he must take the greatest care in a dangerous country to avoid disturbing them. If he sees that he is discovered by some animal or bird, he should immediately stand perfectly still, as this will allay suspicion, while any quick movement or threatening gesture will send it off in a panic.

Game as a rule is less frightened at seeing man than it is on smelling him.

While one whiff of a human being will send a herd of animals off in a wild stampede, they will often, on seeing a man, if he does not move, have a good look at him, and then move off slowly, turning round at intervals, evidently intensely curious.

If a prisoner can be captured, he may be of great use in giving information. He should be interrogated at once, while still under the influence of fear and surprise, and before he has had time to invent any false information.

Questions should be put to him in a haphazard way, and should never be in sequence leading up to the information particularly wanted. One should always assume greater knowledge than one actually possesses. The native's mind should be constantly led away from the issues at stake, returning to them suddenly.

We will suppose that the scout has found some cattle tracks of the day before, and notices one spoor-mark bigger than that of the ordinary native cattle. Soon afterwards, having captured a prisoner, he wishes to find out where the cattle have gone, and what resistance is likely to be made to prevent their capture. This he wishes to report to his column immediately, without wasting a day in following them up, and perhaps in so doing giving the alarm. His questions might be something as follows :

Where are you going to ?
Is there water near here ?
What is your tribe ?
Let me see your tobacco.
Why do you lie to me ? You are not a Swahili.
Why are you at war with us ?
Were you with the cattle I saw down in the plain there yesterday ?
Was it you sitting under the tall tree while they were grazing ?
Do all the cattle belong to the same chief?
Why did you leave this grazing ground ?
What is your name ?
How far have you come to-day ?
Do you know who I am ?
Why do your people call me Khodah ?
Where did the big bull come from, the one with the long horns?
You know very well that it was stolen.
Why have you gone off to such a bad grazing ground?
What is the name of the chief whose village is near the water there where you have gone ?
Are all the cattle that drink there his ?

> *Who is the other big chief with him?*
> *Do all your people file their teeth?*
> *How is this done?*
> *How many men did Chambezi bring with him?*
> *Where do they get their food from?*
> *Where did he leave his women and the rest of his people?*
> *Etc., etc.*

To get really satisfactory information out of a native generally requires great patience. He should be questioned again later, and note how his information agrees with that he gave before. If he was very agitated when he was first captured, he will have forgotten what he really did say, and can be questioned about things he was supposed to have said. Needless to say, different peoples differ enormously in the way they should be dealt with and interrogated.

Reconnoitring by day, proceed cautiously from point to point, always observing the ground to your front and flanks before proceeding farther, and keeping a constant lookout in all directions, and on the ground for spoor of men, horses, and cattle. Never top a rise suddenly, but always proceed slowly, making for some cover on the top from behind which the ground in front can be observed.

If following spoor, it is best if possible not to follow into thick places, but to make a detour so as to cut the outgoing tracks.

Doing this be careful to pass over the best ground possible for observing spoor on. If after circling round no outgoing tracks have been met, it can be assumed that whatever was tracked is still in the cover.

Always go silently, as, in addition to not being heard yourself, you stand a better chance of hearing any one else. Always keep a good look-out for smoke, as this can be seen against a good background for several miles. Remember that by keeping quite still in shadow on ground assimilated to your dress you will practically never be observed, so most of your observations will be performed during the daytime by sitting still in a point of vantage and watching.

If you have any cause to think that you are being followed, you should take steps to ascertain if this is so and throw the pursuer off the track. This can be done by passing over rocky ground and then, if on foot, changing to native shoes or sandals, taking to water, getting on to your old spoor of the morning, and leaving it again in an inconspicuous place, and various other ways which your ingenuity will devise. By going carefully and picking your ground you can at least make the following of your spoor a very slow process. Knowing the habits and customs of the tribes you have to deal with will prevent you from giving yourself away when wearing native footgear.

You should not retrace your steps if it is possible to avoid doing so, as you may be ambushed, and if you wish to strike your old spoor you should avoid thick cover when you do so. It may be useful to be able to distinguish a mob of cattle from horses at a long distance. Horses can be distinguished by having all their heads turned in the same direction while grazing, whereas cattle graze anyhow. Where there were a quantity of horses grazing there would probably be an equal number of men in the close vicinity, whereas with cattle there would probably be only a few men tending them, and they might be some way from a village or the fighting men.

Natives seldom move about at night, and hardly ever singly or in small numbers, having a superstitious dread of being alone in the dark. It will be very seldom that they will make a night attack, although they must be aware of the tremendous advantage the surprise and the approach unseen gives them with their inferior weapons though much superior bushcraft. Some, however, e.g. the Zulus, sometimes attack at or before dawn.

In scouting at night, camp fires and lights should be looked for, sounds listened for, such as drumming, which usually accompanies a war dance, bleating and lowing of stock, talking and singing, dogs barking, and smells sniffed for such as smoke and the dung and scent of stock.

All sounds likely to be heard in the bush should be noticed, so that the ear may become familiar with them. This is more especially useful at night. Knowing these will prevent them being mistaken for anything else, while it will make any strange sound the more noticeable. For instance, some birds whistle exactly like a man attracting attention, and till you are used to it causes you to look round to see who is calling.

Villages or kraals should always be approached upwind, especially at night, as the stock may become restless and any dogs in the village will bark if they get your wind ; moreover you yourself will have the benefit of locating it more accurately by the smell, which at night is a great help. It is also necessary to approach quietly, as dogs are very keen of hearing, especially at night when all is quiet. Dogs of native villages are usually more alert at night and more suspicious of strangers than are those accustomed to civilisation.

Scent is a most elusive thing to follow, as it comes in whiffs and then often cannot be noticed again. One whiff, however, will inform you that there is a smell of fire or stock in the air, and then sounds can be listened for, such as the moving and stamping of animals, or where there are goats the peculiar cry the he-goat makes at night. The sound should be followed for a short distance and then listened for again. It must be remembered that with a cross-wind sound appears to come from a place downwind of where the sound really comes from. When you cross upwind of a village, the farther you are from it at the moment of being exactly upwind from it the better.

If it is necessary to cross upwind and you are sure of your ground with a moderate breeze blowing, six hundred yards' distance should suffice, and half a mile for a strong breeze. When you are at an angle to the wind, how near you can approach depends on the strength of the breeze. With a good breeze which will carry your wind straight past, you might approach within two hundred yards at an angle of 45° upwind from the village, but with a puffy wind this would not do.

If a sentry or post of the enemy is seen and you wish to capture him, watch his movements from a distance for a time to see if he has any one near him or any supports. If you can circle round behind him, any tracks would give you the necessary information. To do this, or to get to a better spot to watch him from, you must mark down his position (generally by means of trees) and your best method of approach under cover. When you have noted landmarks on the route to be taken and committed the country to memory, slowly retire out of view and make your stalk. The stalk should be made without again coming into view. The sun and direction of the wind should be carefully noticed to guide you by. If you are mounted and he is on foot, a rush would generally get him, provided the country be not too thick. If he is mounted, endeavour to capture, shoot, or even frighten away his horse. You should get, if possible, on his probable line of retreat, which will cause him to be undecided in which direction to go.

During the last stages of the stalk your native tracker might appear from the opposite side, to distract his attention and possibly cause him to retreat your way. When he sees you, he will probably think that he is surrounded on all sides and that it is no use running away.

APPENDIX I.

HOW TO USE THE STAR CHARTS.

TAKE the required chart and look for the time of the night. Follow this column, round (or down, in the equatorial chart), till you come to your month.

If you are nearer the 1st of the month, take the line given against this time and date, but if you are nearer the 15th of the month, you must take the line 1 hour *later* than it really is.

This line then gives the part of the heavens which have reached the highest point in their revolution, at the time of observation.

Having got this line, if you are looking at the chart of the Northern Constellations, face north and hold up this line perpendicularly (date and time uppermost).

The chart is then " set."

If you were standing on the equator, this line would then represent an imaginary line, drawn from the point of the heavens exactly over your head, to the

North Pole of the heavens on the horizon due north of you; and the top half of the chart would represent the stars then above the horizon in the Northern Hemisphere, while the bottom half would be those below the horizon.

If you are north of the equator, the farther north you are the higher will the North Pole of the heavens be above the horizon and the more will you be able to see of the lower half of the chart.

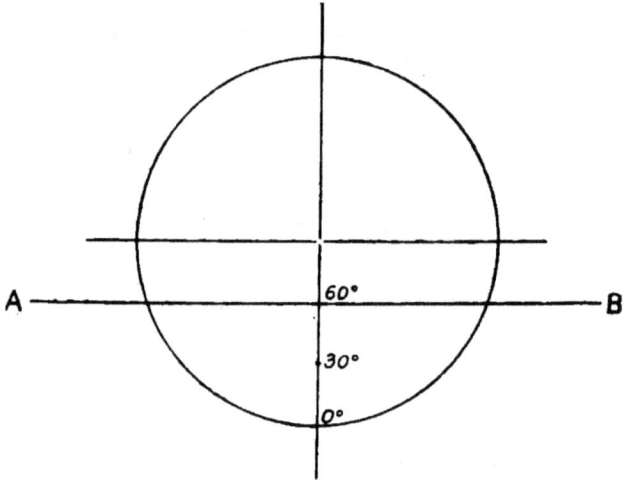

Thus if you were in lat. 30° N. and could see the horizon all round, you ought to be able to see 30° below the pole, or a third of the way down the lower half of the chart—that is, to the 60° circle.

Thus in lat. 30° N. you should see all the stars that are, speaking roughly, above line AB in the diagram. In practice, of course, you will seldom be able to see stars close down on the horizon.

If you are south of the equator, conversely, you see less and less of the northern stars.

To use the chart of the Southern Constellations, you proceed exactly in the same way, only *facing south*.

For the Equatorial Chart, find your line for time and date as before.

Then for eastern stars face east.

Count six lines *below* your line, and between these two lines gives the celestial equator from the horizon to its highest point.

Tilt this equator line to correspond with your latitude—that is, if you are south of equator, you must tilt the top of the line to the left the amount of your latitude, and if you are north *vice versa*.

The chart is then set, and the rest of the chart can be folded up.

E.g. Your lat. is 45° N.

You tilt the top of your equatorial line half a right angle to the right.

Then a horizontal line through the equator, where it is cut by the sixth line below, will roughly give you your horizon—that is to say, that some stars on the left, which would be below the horizon if the chart were held perpendicularly, are now brought above, and some on the right have been sent below.

For western stars face the west, but with the chart *upside down*. Count six lines above your line (that is, downwards with the map upside-down) and proceed as below.

Conversely, if the time is unknown, from a

knowledge of the equatorial stars and the approximate date, the time can be judged at night to within half an hour, as those stars on and quite close to the equator rise six hours before they reach their highest point and set six hours after reaching their highest point.

Each transverse division across the equatorial map represents an hour in time.

E.g. On November 1st you see Procyon rising.

Referring to the Equatorial Chart, you see that Procyon is distant from the line above it about twice as much as that below it.

The line above shows 4 a.m. for November 1st, and that below 5 a.m. for same date. Therefore Procyon will be at its highest point two-thirds of an hour more than 4 a.m., *viz.* 4.40 a.m., and so will rise six hours earlier, *viz.* 10.40 p.m.

If you wish to use any star for bearings other than those on pages 19, 20 and 21, find them on the Northern or Southern Constellation chart (as the case may be), and by the circles and degrees marked on the charts its declination can be judged within a degree or so, and having obtained this use the star as directed on page 21 and following pages.

APPENDIX II.

EXERCISES FOR SCOUTS.

KNOWLEDGE of country may be divided into two branches, which might be compared to strategy and tactics respectively. The strategy of knowledge of country consists in finding one's way about large areas of wild country. This necessarily can only be practised in special countries. The tactics, however, which consists of reconnoitring, marking down an enemy, and stalking him, can be practised in most places.

To practise stalking and approaching under cover, post a man on a hill or other look-out station with a rifle and blank ammunition, and endeavour to approach him from a distance unseen. Meanwhile, he is to fire a round every time a fair shot is offered. Post a man in a wood and make another look for him and endeavour to locate him unseen. Make two scouts start at different ends of a path or road, and see which can see the other first. Mark down some spot in a wood or on a heath which has no road, path, or hedge near it, and try to walk straight to it some days after. Mark down some tree or bush, and, retiring from it, endeavour to make a detour and come up to it from another direction, without coming in sight of it again till you are quite close.

Whenever you take a walk, try to find your way back by a different route without reference to signposts.

Numberless other exercises of a similar nature can be devised.

Quickness of perception and observancy of detail can be practised in thousands of ways even in a crowded city. Two scouts can have a match as to which of them can see certain things quickest as they walk along. Marks can be allotted to the one that sees a thing first, and a hundred up might be game. These things might vary from time to time, but there should always be some things on the ground, so as to practise looking on the ground for spoor, and there should be moving objects, so as to have to search the horizon at the same time. Such things might be, in a town:

Match ends.	1 mark.
Fusee ends.	3 marks.
Lost buttons.	4 marks.
Lost hairpins.	1 mark.
Men wearing panama hats.	1 mark.
Fox terriers.	2 marks.

While another day cigar-ends, beads or

spangles, matchboxes and feathers on the ground, and bath-chairs and cats in the distance might be looked for. So as to practise listening for sounds as well, marks might also be allotted for hearing cats or recognising the sound made by an omnibus.

In the country more realistic work may be done, such as awarding marks for spoor of cattle, nonslipping heel-pads, spoor of pigs, hearing dogs barking, seeing rooks disturbed by the presence of man, seeing horses, and smelling fires. Any exercise which makes one notice detail helps to quicken the perception.

Passing a shop window, try and recall afterwards as many of the things seen as possible, or endeavour to write down everything seen in a room, its position, and what deductions were made, from it.

Walking along in the country, train the eye to pick up distant objects, and at once catch sight of anything moving in any direction.

If a bird is heard, try to locate its position at once. Practise describing the position of places, routes, etc. It is most uncommon to meet any one who can give you clear and accurate directions of how to find a place or where to look for an object.

Practise this pointing out of objects quickly to some one else, as also the locating of objects pointed out to you. When you return from a march or walk, try to recall everything seen and noted on the way, including useful landmarks, bearings, by-paths, and other routes, and where you think they lead to, etc.

Scouts might compete in writing out all they have seen and noticed in a day's exercise.

Practise in judging distance must be carried out frequently. Distances can be checked from a map when you are standing in a known position. Write down the distances you estimate to different points from an elevated position, then check by a map and look at them again. Or the instructor can work them out from the map beforehand, and all he has to do is to recognise them on the ground.

Practise bearings and time from sun and stars. This you can be always noticing whenever any are visible.

For practising tracking, commence on the sands or a lonely dusty road, where all tracks can be seen easily, so as to get familiar with the look of different tracks, moving at different rates of speed. After this try to distinguish any flaws or peculiarities by which individual tracks may be recognised. For the beginner, tracks of different men and animals at different paces might be made side by side and at different intervals of time for comparison.

When these have been mastered, the scout must be made to guess the age of spoors, and say all that he can find out from them. As an example of how to look at a track and make deductions from it, we will imagine that the track of a bicycle is under consideration.

First to tell in which way it is going. This may be told by—

(1)   On an incline, a zigzag track in ascending and a straight track in descending.

(2)   Towards which way a rut passed over crumbles.

(3)   A stone or bit of grit turned over or pushed forward, leaving either the mark of its original impress behind it or the track of where it has been pushed along.

(4)   When the bicycle passes through a pool of mud or through a muddy patch of the road it will be seen to leave a dry spoor where going in and a wet spoor coming out. Similarly it can be seen which side of a dusty patch the spoor shows most dust, or when the bicycle is wheeled on and off the grass it will show a dirty track where wheeled on and a clean one where wheeled off

(5)   In wet weather splotches of mud off the wheels, splashing forward in the direction in which the cycle is going.

(6)   Which side of the road it is going.

Next the pace. It can be told roughly whether it was going fast or slow by the amount of wobble it has, and also if mud splashes have fallen in large or very flat splashes. The faster the pace the greater is the force with which a splotch of mud hits the ground, and hence the farther forward it splashes. A very sharp turn can only be made when going very slowly.

Individual bicycles can be told by noticing the kind of tyre, and then finding some flaw or scratch in it which leaves its impress in the spoor. It must be remembered that after passing over stony ground new flaws may appear. If it is thought that the wheel differs from the usual size, its circumference can be ascertained by measuring from the mark where a flaw occurs to where that same mark occurs the next time. Where the brake has been put on can sometimes be seen by the splashing of mud to the sides.

A cyclist can be told when not with his machine by having his trousers frayed or else creased where they have been fastened up, by having specks of mud on his back, by having the soles of his boots or shoes scored by the pedals. The latter might be seen in his walking spoor.

The marks made by axes on trees at different intervals of time should be compared, and the scout should be made to guess the times that have elapsed since different leaves and shoots were trodden down or plucked.

Parties of scouts should be taken out and made to make a rough sketch of their route on return from notes made on the way. Map reading should also be constantly practised.

APPENDIX III.

STAR CHARTS.

(1.) Northern Hemisphere Star Chart

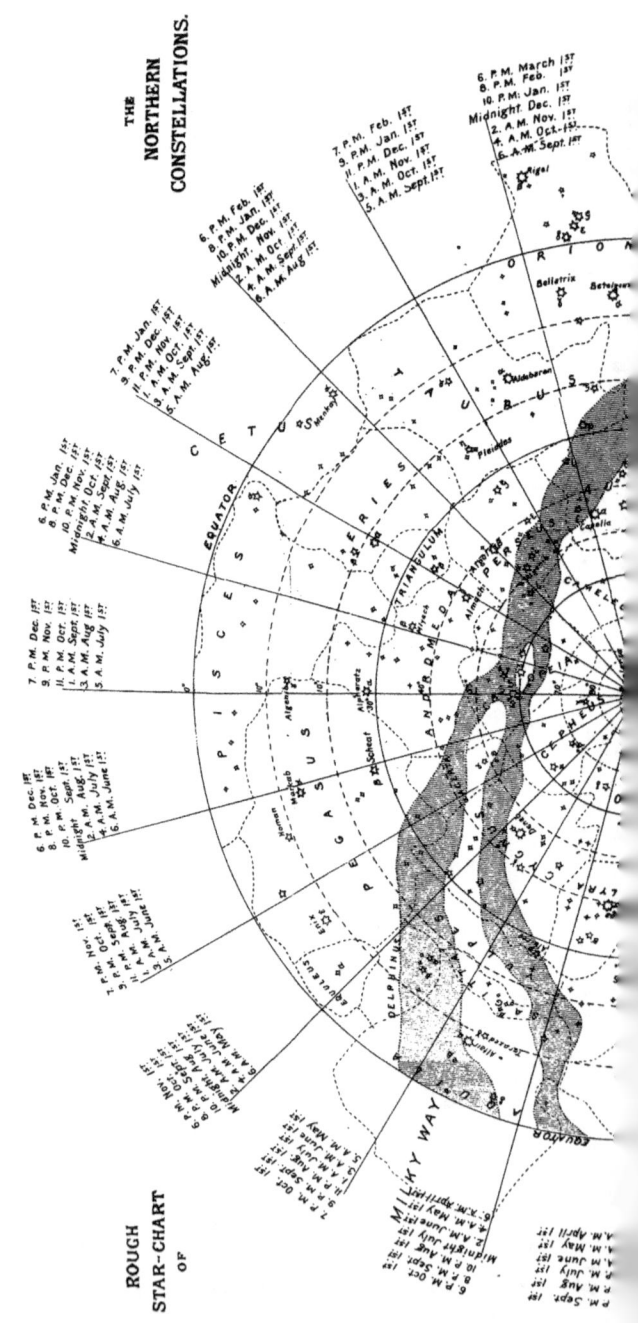

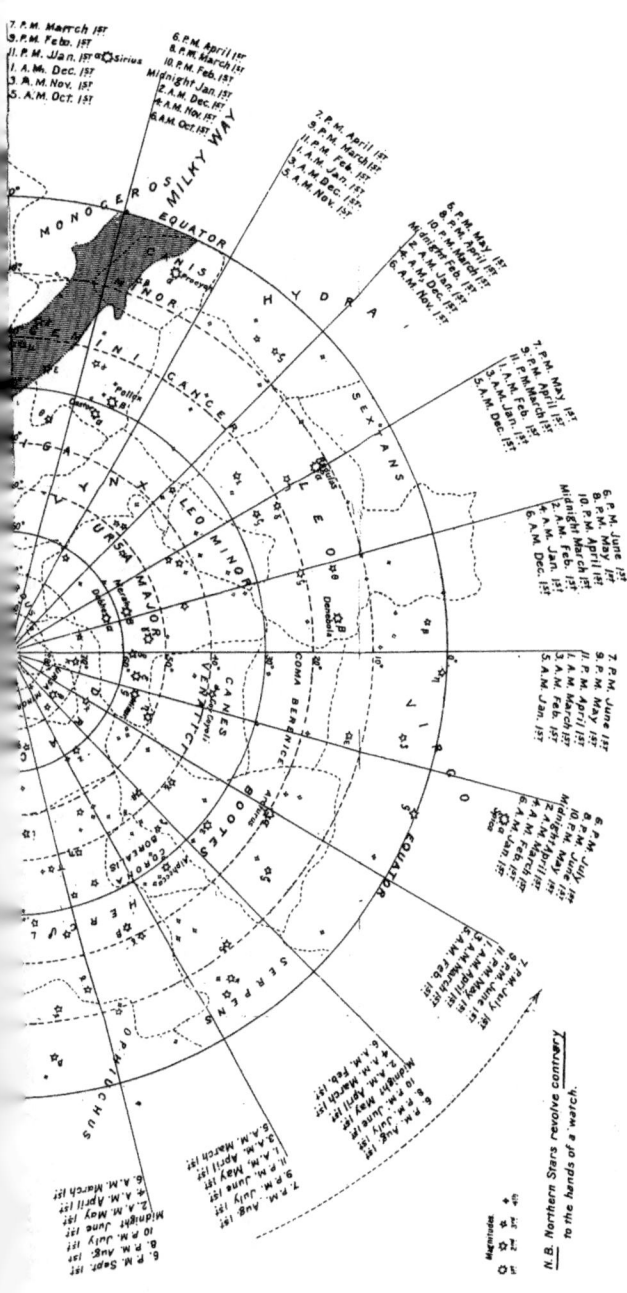

(2.) Southern Hemisphere Star Chart

122 SCOUTING & RECONNAISSANCE

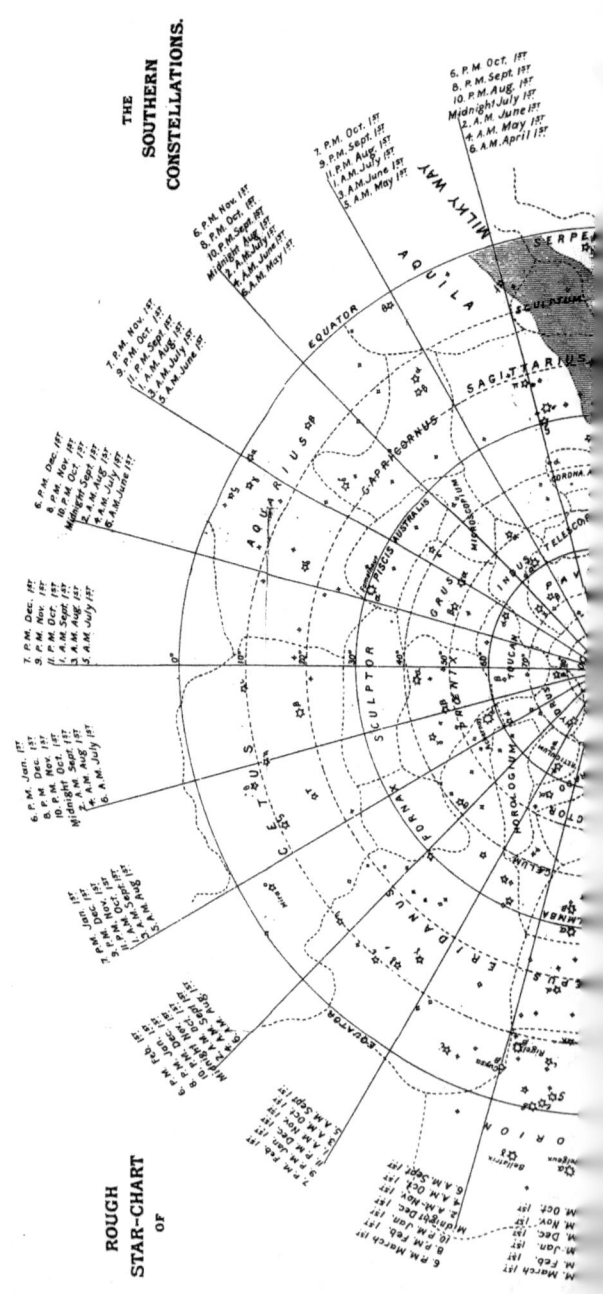

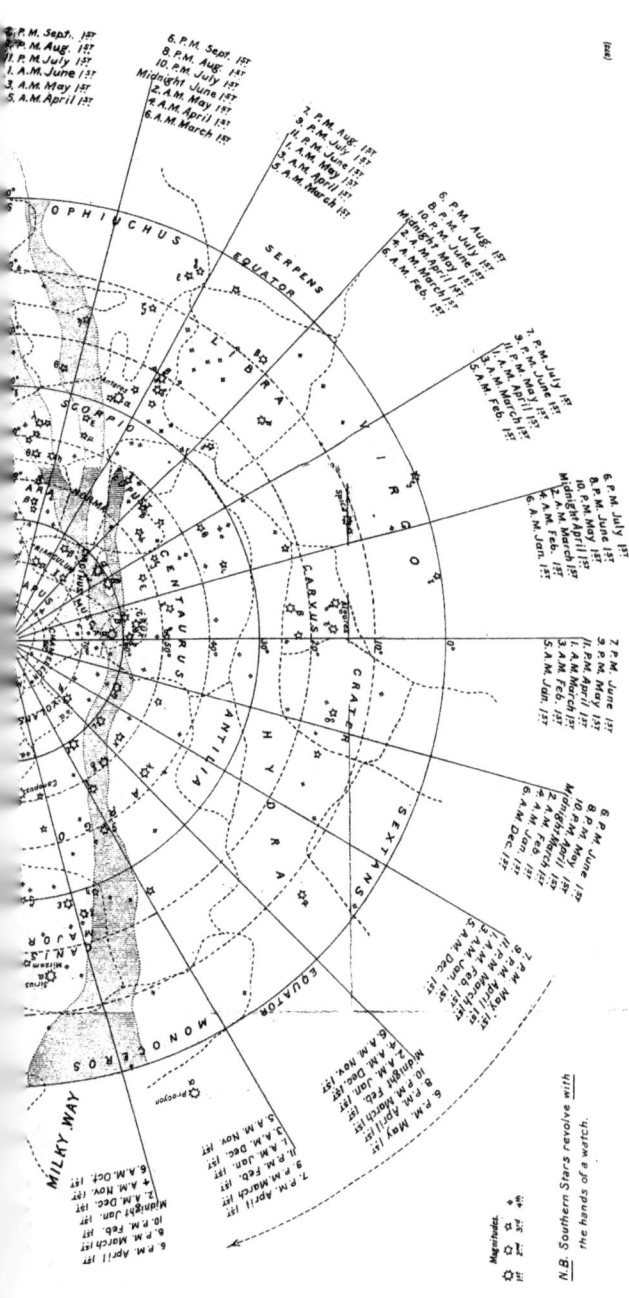

(3.) Equatorial Star Chart

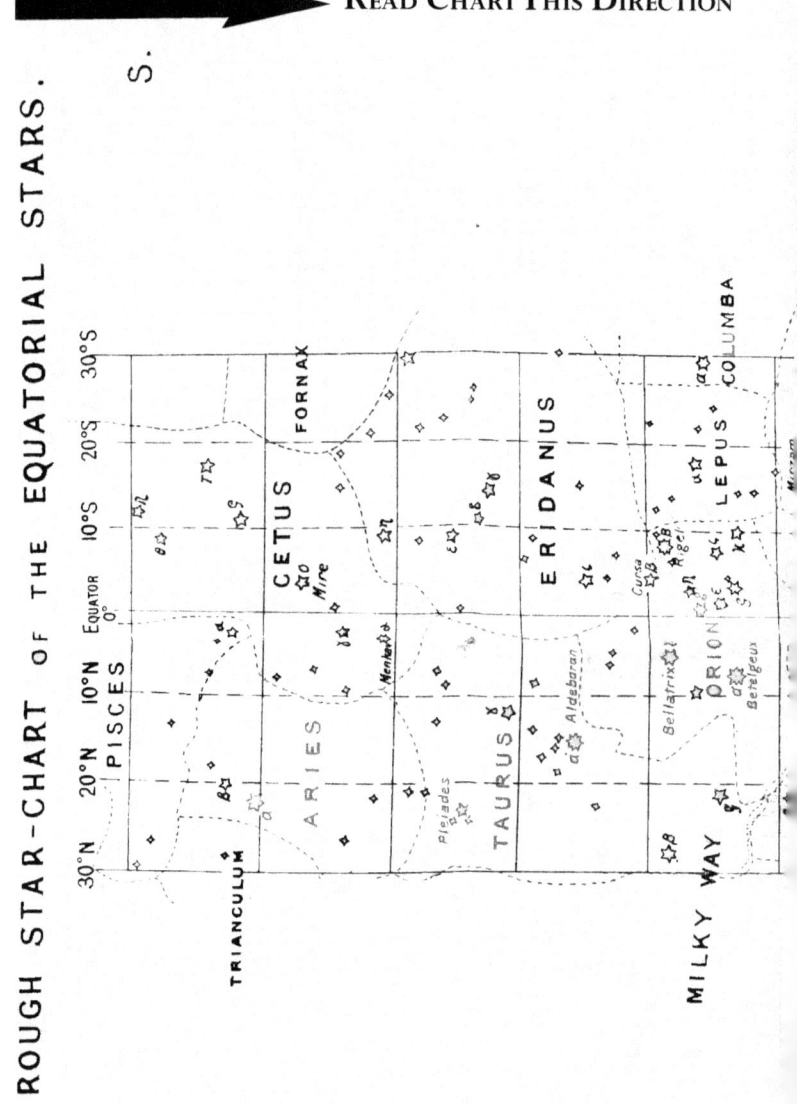

IN SAVAGE COUNTRIES   127

SCOUTING & RECONNAISSANCE

READ CHART THIS DIRECTION →

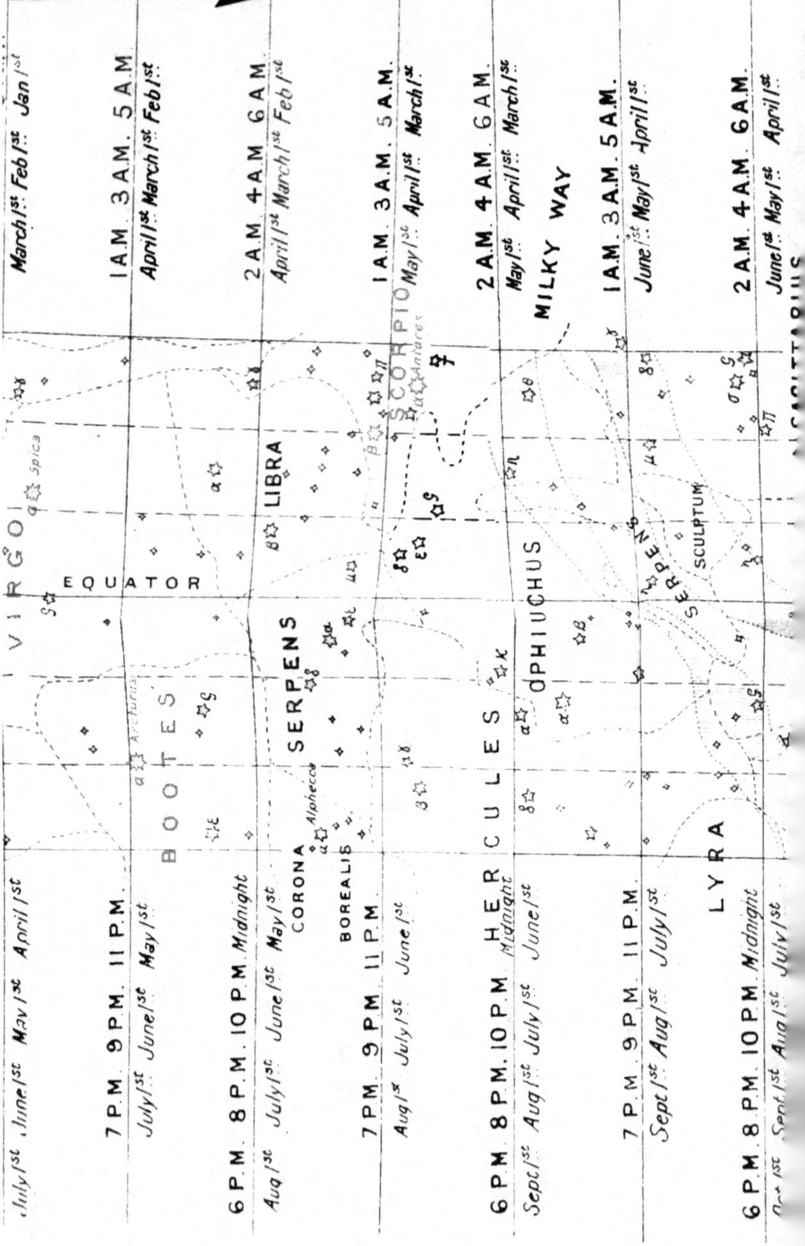

(Star chart showing constellations including VIRGO, BOOTES, SERPENS, CORONA BOREALIS, LIBRA, SCORPIO (with Antares), HERCULES, OPHIUCHUS, SCULPTUM, LYRA, SAGITTARIUS, with the EQUATOR and MILKY WAY marked.)

Top time/date labels (left to right):
- March 1st :: Feb 1st :: Jan 1st
- 1 A.M. 3 A.M. 5 A.M. — April 1st :: March 1st :: Feb 1st
- 2 A.M. 4 A.M. 6 A.M. — April 1st :: March 1st :: Feb 1st
- 1 A.M. 3 A.M. 5 A.M. — May 1st :: April 1st :: March 1st
- 2 A.M. 4 A.M. 6 A.M. — May 1st :: April 1st :: March 1st
- 1 A.M. 3 A.M. 5 A.M. — June 1st :: May 1st :: April 1st
- 2 A.M. 4 A.M. 6 A.M. — June 1st :: May 1st :: April 1st

Bottom time/date labels (left to right):
- July 1st :: June 1st :: May 1st :: April 1st
- 7 P.M. 9 P.M. 11 P.M. — July 1st :: June 1st :: May 1st
- 6 P.M. 8 P.M. 10 P.M. Midnight — Aug 1st :: July 1st :: June 1st :: May 1st
- 7 P.M. 9 P.M. 11 P.M. — Aug 1st :: July 1st :: June 1st
- 6 P.M. 8 P.M. 10 P.M. Midnight — Sept 1st :: Aug 1st :: July 1st :: June 1st
- 7 P.M. 9 P.M. 11 P.M. — Sept 1st :: Aug 1st :: July 1st
- 6 P.M. 8 P.M. 10 P.M. Midnight — Oct 1st :: Sept 1st :: Aug 1st :: July 1st

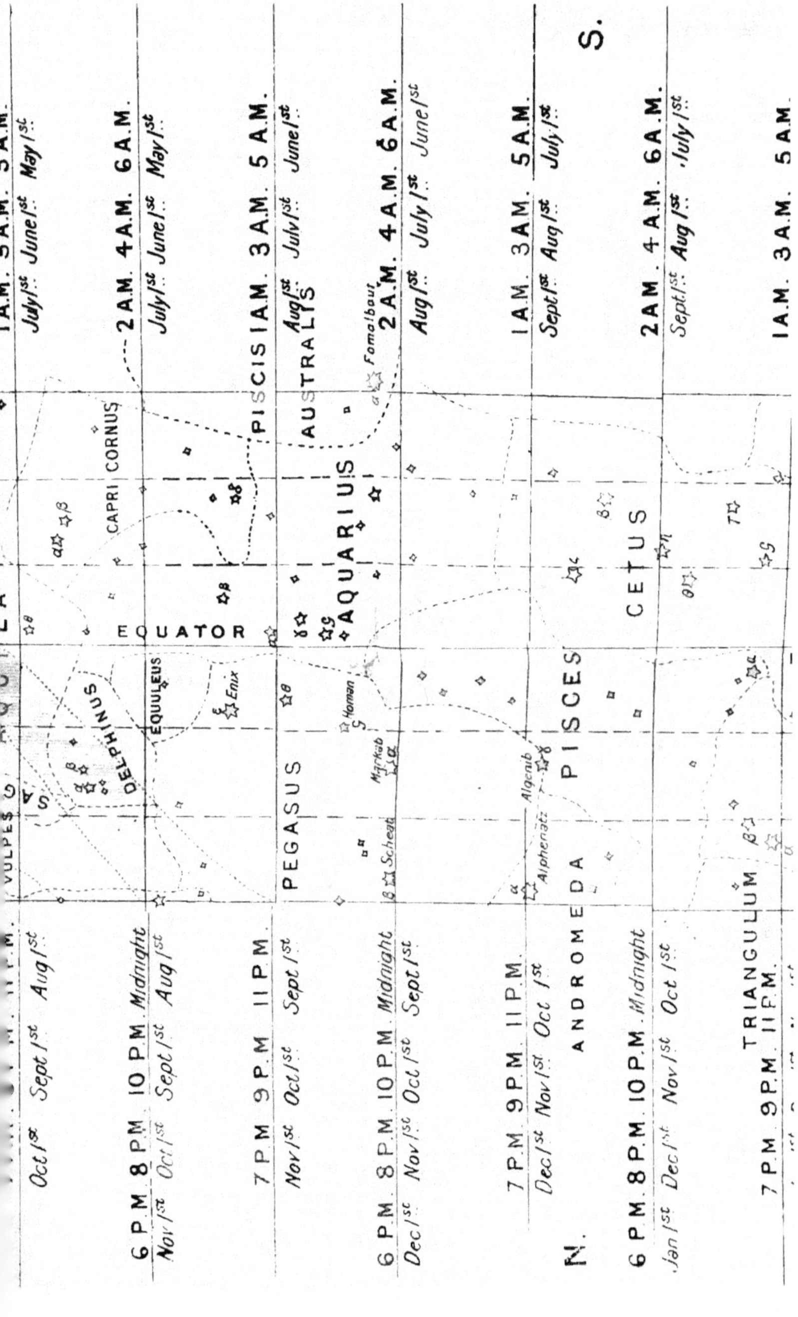

IN SAVAGE COUNTRIES

[Star chart showing southern sky constellations including Aquila, Vulpecula, Delphinus, Equuleus, Pegasus, Capricornus, Aquarius, Piscis Australis, Pisces, Andromeda, Cetus, and Triangulum. Time and date markings around the edges indicate viewing times from 6 P.M. through 6 A.M. across various months.]

Top edge (right side, reading down):
- 1 A.M. 3 A.M. 5 A.M. — July 1st, June 1st, May 1st
- 2 A.M. 4 A.M. 6 A.M. — July 1st, June 1st, May 1st
- PISCIS 1 A.M. 3 A.M. 5 A.M. — Aug 1st, July 1st, June 1st
- AUSTRALIS — α Fomalhaut
- 2 A.M. 4 A.M. 6 A.M. — Aug 1st, July 1st, June 1st
- 1 A.M. 3 A.M. 5 A.M. — Sept 1st, Aug 1st, July 1st
- 2 A.M. 4 A.M. 6 A.M. — Sept 1st, Aug 1st, July 1st
- 1 A.M. 3 A.M. 5 A.M.

S.

Constellations visible: VULPES, AQUILA, DELPHINUS, EQUULEUS, Enix, PEGASUS (Homan, Markab, Scheat), CAPRICORNUS, EQUATOR, AQUARIUS, PISCES (Algenib, Alpheratz), ANDROMEDA, CETUS, TRIANGULUM

N.

Bottom edge (reading down):
- Oct 1st, Sept 1st, Aug 1st
- 6 P.M. 8 P.M. 10 P.M. Midnight — Nov 1st, Oct 1st, Sept 1st, Aug 1st
- 7 P.M. 9 P.M. 11 P.M. — Nov 1st, Oct 1st, Sept 1st
- 6 P.M. 8 P.M. 10 P.M. Midnight — Dec 1st, Nov 1st, Oct 1st, Sept 1st
- 7 P.M. 9 P.M. 11 P.M. — Dec 1st, Nov 1st, Oct 1st
- 6 P.M. 8 P.M. 10 P.M. Midnight — Jan 1st, Dec 1st, Nov 1st, Oct 1st
- 7 P.M. 9 P.M. 11 P.M.

130   SCOUTING & RECONNAISSANCE

Read Chart This Direction →

S

1 A.M. 3 A.M. 5 A.M.	2 A.M. 4 A.M. 6 A.M.	PISCIS 1 A.M. 3 A.M. 5 A.M.	AUSTRALIS 2 A.M. 4 A.M. 6 A.M.	1 A.M. 3 A.M. 5 A.M.	2 A.M. 4 A.M. 6 A.M.
July 1st.. June 1st.. May 1st	July 1st.. June 1st.. May 1st	Aug 1st.. July 1st.. June 1st	Aug 1st.. July 1st.. June 1st	Sept 1st.. Aug 1st.. July 1st	Sept 1st.. Aug 1st.. July 1st

CAPRICORNUS ... AQUARIUS ... Fomalhaut ... CETUS

AQUILA ... DELPHINUS ... EQUULEUS ... Enix ... EQUATOR

VULPES ... PEGASUS ... Markab ... Scheat ... Homan ... Algenib ... Alpheratz ... PISCES ... ANDROMEDA ... TRIANGULUM

7 P.M. 9 P.M. 11 P.M.	6 P.M. 8 P.M. 10 P.M. Midnight	7 P.M. 9 P.M. 11 P.M.	6 P.M. 8 P.M. 10 P.M. Midnight	7 P.M. 9 P.M. 11 P.M.	6 P.M. 8 P.M. 10 P.M. Midnight
Oct 1st.. Sept 1st.. Aug 1st	Nov 1st.. Oct 1st.. Sept 1st.. Aug 1st	Nov 1st.. Oct 1st.. Sept 1st	Dec 1st.. Nov 1st.. Oct 1st.. Sept 1st	Dec 1st.. Nov 1st.. Oct 1st	Jan 1st.. Dec 1st.. Nov 1st.. Oct 1st

N.

IN SAVAGE COUNTRIES 131

		2 A.M.	4 A.M. 6 A.M.
		Oct 1st	Sept 1st Aug 1st
	1 A.M.	3 A.M.	5 A.M.
	Nov 1st	Oct 1st	Sept 1st
		2 A.M.	4 A.M. 6 A.M.
		Nov 1st	Oct 1st Sept 1st
	1 A.M.	3 A.M.	5 A.M.
	Dec 1st	Nov 1st	Oct 1st

FORNAX

Mira

—ATOR Menkar η α

TAURUS δ
Pleiades α ☆ Aldebaran

ε ☆ δ γ

ERIDANUS
ι
β Cursa
λ Rigel
ORION
Bellatrix γ
α Betelgeux

LEPUS α ☆
κ ☆ COLUMBA

β ☆ Mirzam
α ☆ Sirius
CANIS
MAJOR
ε ☆
η ☆

MONOCEROS

CANIS
MINOR
β
α Procyon

GEMINI
β ☆ Pollux
α ☆ Castor

CANCER

MILKY WAY

A R G O

6 P.M. 8 P.M. 10 P.M. Midnight
Feb 1st Jan 1st Dec 1st Nov 1st

7 P.M. 9 P.M. 11 P.M.
Feb 1st Jan 1st Dec 1st

6 P.M. 8 P.M. 10 P.M. Midnight
March 1st Feb 1st Jan 1st Dec 1st

MILKY WAY

7 P.M. 9 P.M. 11 P.M.
March 1st Feb 1st Jan 1st

132 SCOUTING & RECONNAISSANCE

 **Read Chart This Direction**

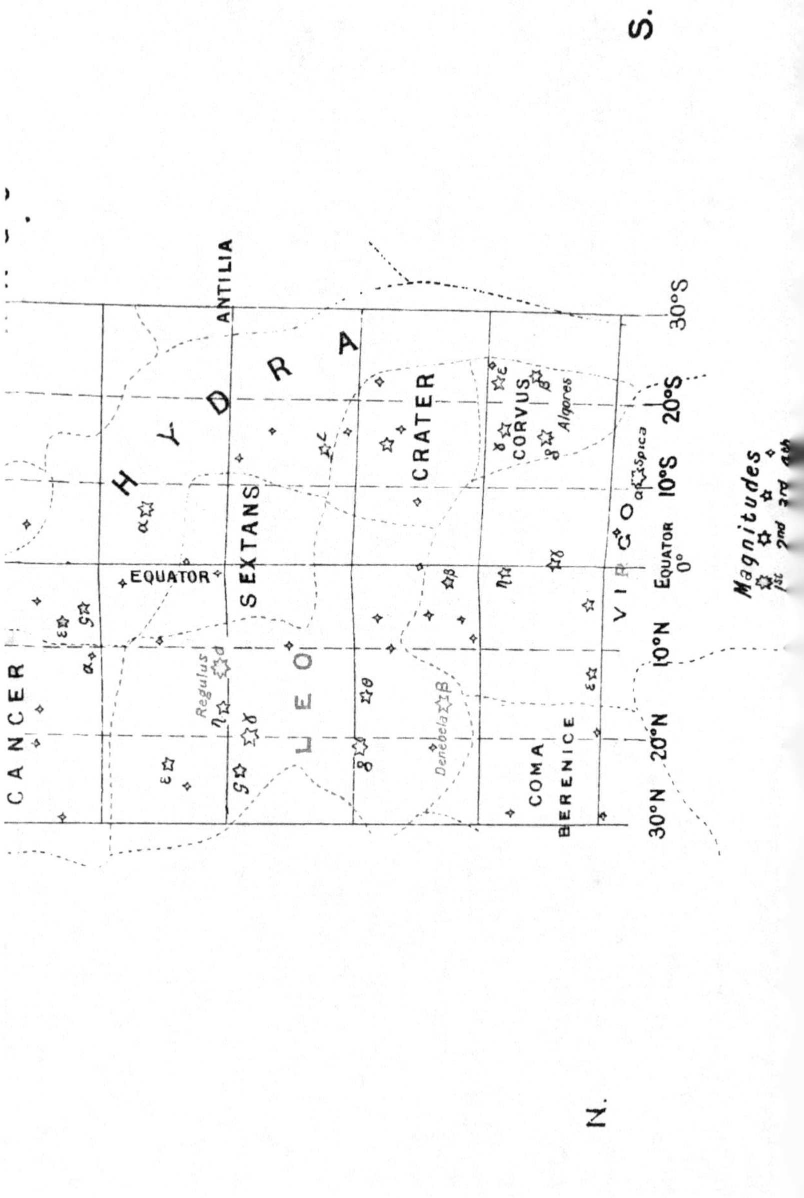

N.B. North is to the left and South to the right.

Hour. 6 P.M to Midnight on the left and 1 A.M to 6 A.M. on the right.

INDEX.

	Page
Apparent time and mean time	25
Arab, spoor of	62
Axe for a scouting	67
Base, measuring a	52
Bearings, errors from marching on	33
Bearings, from moon	12
Bearings, from stars	12
Bearings, from sun	3
Bearings, from wind	22
Bearings, how obtained	2
Bearings, must be supplemented by landmarks	48
Bearings, of sun at rising and setting	10
Bearings, practice judging, at all times	112
Bearings, use true and not magnetic	50
Biltong	76
Bivouac, how to	78
Bivouacking, method of judging when to march at night	31
Blanket for mounted scout	77
Blazing trees	85
Blood spoor	85
Boots should be constantly greased	82
Browsing of different animals	63
Browsing, tracking by	63

Bush paths, how to traverse 41
Bush paths, to distant places 41
Bushmen's signs 84

Camel, difference of individual tracks 65
Camel, spoor of 65
Canteen for scout 76
Caste marks 89
Cattle, how to distinguish, in distance 100
Cattle, spoor of 62
Cattle, spoor of, on trek 63
Centaurus, obtaining south from pointers of 14
Charts, star 115
Charts, star, how to use 105
Chill, how to avoid, at night 79
Clothing for invisibility 71
Clothing, kind suggested 72
Compass, checking variation of 12
Compass, for scout 75
Compass, may be impossible to use a 3
Cooking meat when bivouacked 79
Customs and languages of natives must be learnt 87

Declination of sun explained 4
Declination of sun, table of 5
Declinations of stars, northern hemisphere 15
Declinations of stars, southern hemisphere 16
Desert country 51
Determining age of spoor by drinking places 67
Determining age of spoor by droppings 68
Determining breadth of marsh 53
Direction, means of finding your 2
Direction, obtained differently in flat and hilly country 40
Direction of an object, pointing out 111
Direction, when landmarks alone will not suffice 43

Directions, difference in, by plainsman and hillsman 93
Directions, practise pointing out 111
Distance traversed, means of judging 112
Dog, spoor of 66
Donkey, spoor of 65
Droppings 68
Droppings, age of spoor from 68
Dung as landmarks, animals' 42

Equinox, time of sunrise at 25
Equipment for scout 72
Errors in marching on bearings 33
Fetish animals 90
Field-book, example of rough 47
Field-book, example of how it may serve 46
Field-book, example of notes accompanying 46
Field-book should be made 45
Field-glasses for scout 75
Field-glasses, need of practice with 76
Filter for muddy pools 78
Finding a place fixed by one distant landmark 44
Finding way in unknown country 44
Fire, how to feed, at night 80
Fire, native methods of making 91
Fires at night in hostile country 78
Fires, look for, when scouting 100
Fixing a place by distant landmarks 43
Food, customs connected with 89
Footgear, native 62

Game paths 42
Game paths use of 42
Geology, rough idea of, useful 77
Ghurka signs 92
Goat, spoor of 64

Grain for horse 77

Handicrafts, native 91
Heat haze, effect of 52
Hill, rough sketch to remember shape 35
Hoofed animals, cutting grass and leaves 60
Hoofed animals, spoor of 57
Horses, how to distinguish in distance 100
Horses, spoor of 64

Indian, spoor of 62

Judging distance 112

Knife for scout 75
Knowledge of country, definition of 1
Landmarks 35
Landmarks, animals' dung as 42
Landmarks, fixing a place by distant 42
Landmarks, hills as 35
Landmarks, in rear, notice carefully 40
Landmarks, needles, perched blocks, etc. 40
Landmarks, outcrop of rock as 2
Landmarks, overjudging lateral distance from 49
Landmarks, slight rises in wooded country form 50
Landmarks, the different kinds used 2
Landmarks, to check bearings by sun, etc. 40
Landmarks, tracks and watercourses 39
Landmarks, trees as 37
Landmarks, white ant-hills as 43
Languages and customs must be learnt 87
Latitude, how obtained approximately 6
Lengths of shadow and how to calculate them 9
Longest and shortest days 26
Man, detecting presence of 95

Map-reading to be practised 114
Marching at night, how to tell hour to start 32
Marching at night, on a star 20
Marsh, determining breadth of 53
Matches, how to carry 75
Mean time and apparent time 25
Measuring a base 52
Moon, bearings from 12
Moon, declination of 12
Moon, movements intricate 12
Moon, rises and sets due E. and W. twice a month 12
Moon, time by 30
Movements of sun described 3

Native footgear 62
Native tracker 76
Native tracks. See *Bush paths*
Native, unreliable guide at night 93
Negro, spoor of 61
Night attacks by natives 100
Noonday shadow 28
Noonday shadow, how to find length of 28
North, obtaining direction of 9
Note-book for scout 75

Orion for finding E. and W. 15
Outcrops of rock 39

Paths, bush, how to take general bearing of 41
Paths, bush, to distant places 41
Paths, game 42
Pig, spoor of 65
Pointers of Ursa Major, when upright 31
Pointers of Centaurus, south from 14
Pole star, height above horizon 13

Pole star, obtaining north from 12
Pole star, where visible 13
Poles of heavens 12
Position of place or object, describing 81
Precautions before camping in hostile country 78
Prisoner, how to capture a 102
Prisoner, how to interrogate a 96

Quickness of perception, practices for 110

Ration, emergency, for scout 75
Ration, food, for scout 76
Reconnoitring in thick cover 50
Revolver at night 75
Rifle at night 75
Rifle for scout 74
Rock, outcrops of 39
Rope to hobble horse 77

Scout, equipment for 73
Shadow at equinox 9
Shadow at noon 9
Shadow at times other than equinox 9
Shadow, how to find length of, during day 10
Shadow, how to find length of, during noonday 27
Shadow, keep in, reconnoitring 99
Shadows to be watched for bearings 10
Sheep, spoor of 63
Shooting a man 74
Shooting a dying animal 74
Shortest and longest days 26
Signals, arrange with native tracker 84
Signs, bushmen's 84
Signs of prevailing winds 22
Signs, to mark route through dense undergrowth 85

Silence, importance of, in reconnoitring 99
Smelling for stock 95
Smoke, keep look-out for 99
Smoke, not so noticeable in trees 79
Smoking, customs connected with 89
Solstices explained 7
Solstices table of sun's bearing at 8
Somali, spoor of 62
Sore feet 82
Sound, rate at which it travels 52
Sounds to be listened for while scouting 100
South, how to obtain 14
Southern Cross, obtaining south from 14
Southern Cross, when upright 30
Spoor, customs of people spoored must be known 68
Spoor, determining age of 60
Spoor, determining age of from drinking places 67
Spoor, difference between fore and hind 64
Spoor, difference between man and woman 61
Spoor, how to conceal your 79
Spoor, how to look for 56
Spoor, in different soils 56
Spoor, encampments and bivouacs 68
Spoor, grass 58
Spoor of barefooted man 61
Spoor of booted man 62
Spoor of camel 65
Spoor of cattle 62
Spoor of different races 61
Spoor of dog 66
Spoor of donkey 65
Spoor of goat 64
Spoor of hoofed animals 57
Spoor of horses 64
Spoor of sheep 63

Spoor of vehicles 67
Spoor of wild animals 66
Spoor, other signs 67
Stalking an enemy 109
Stalking, how to practise 109
Star charts 115
Star charts, how to use 105
Stars, bearings from 12
Stars, declinations of, in northern hemisphere 15
Stars, decinations of in southern hemisphere 16
Stars, estimating lateral movement of 19
Stars, estimating time above horizon of 19
Stars, places of rising and setting of 19
Stars, rough allowance for lateral movement of 19
Stars, time by 30
Sun at solstices, table of bearings of 8
Sun, bearings from 3
Sun, declination of, explained 4
Sun, movement of, during equinoxes 7
Sun, movements of 7
Sun, table of declinations of 10
Sun, time by 25
Sunrise, allowance for time of 27
Sunrise, ascertaining time of 26

Telescope 76
Time by moon 30
Time by stars 30
Time by sun 25
Time, during day by sun 25
Time of sunrise, how ascertained 26
Tracking a bicycle 112
Tracking by browsing 63
Tracking by dew and rain 59
Tracking by grazing 60
Tracking, constant practice required 112

Tracking, leaves and grass cut by hoofed animals 57
Tracking, on beaten road 59
Tracking, practice for 112
Tracking, what it consists of 55
Tracks, return so as to cut your outgoing 39
Tracks, see *Spoor* and also *Bush paths*

Traversing bush paths 53
Trees as landmarks 37
Trees, blazing 85
Trees, characteristic 38
Trees, importance of knowledge of 51
Trees, importance of learning names of 35
Trees, sketch important 37
Trees, use of belt of 37
Tribal marks 89
Tropics, limits of 4

Unknown country, finding way in 44
Unknown country, what to notice in 44

Vegetables, native, should be found out 80
Vehicles, tracks of 67
Venus, use of 32
Villages, approaching at night 101

Watch for scout 75
Watch, set at sunrise 25
Watch, value of being independent of 25
Water for scout 76
Water, how to find 45
Weapons, native 90
White ant-hills as landmarks 43
Wind, bearings from 22
Wind, methods of noting direction of 23
Wind, traces left by 24

Winds recurring yearly 23
Wind, signs of prevailing 24
Wooded country without landmarks 51

www.ingramcontent.com/pod-product-compliance
Lightning Source LLC
Chambersburg PA
CBHW070758020526
44118CB00036B/1951